Practicing Relativism in the Anthropocene

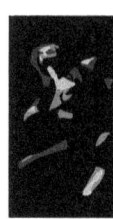 # Critical Climate Change

Series Editors: Tom Cohen and Claire Colebrook

The era of climate change involves the mutation of systems beyond 20th century anthropomorphic models and has stood, until recently, outside representation or address. Understood in a broad and critical sense, climate change concerns material agencies that impact on biomass and energy, erased borders and microbial invention, geological and nanographic time, and extinction events. The possibility of extinction has always been a latent figure in textual production and archives; but the current sense of depletion, decay, mutation and exhaustion calls for new modes of address, new styles of publishing and authoring, and new formats and speeds of distribution. As the pressures and re-alignments of this re-arrangement occur, so must the critical languages and conceptual templates, political premises and definitions of 'life.' There is a particular need to publish in timely fashion experimental monographs that redefine the boundaries of disciplinary fields, rhetorical invasions, the interface of conceptual and scientific languages, and geomorphic and geopolitical interventions. Critical Climate Change is oriented, in this general manner, toward the epistemo-political mutations that correspond to the temporalities of terrestrial mutation.

Practicing Relativism in the Anthropocene
On Science, Belief, and the Humanities

Barbara Herrnstein Smith

O
OPEN HUMANITIES PRESS
London 2018

First edition published by OPEN HUMANITIES PRESS 2018

Copyright © 2018 Barbara Herrnstein Smith

Freely available online at:
http://openhumanitiespress.org/books/titles/practicing-relativism-in-the-anthropocene

This is an open access book, licensed under Creative Commons By Attribution Share Alike license. Under this license, no permission is required from the authors or the publisher for anyone to download, reuse, reprint, modify, distribute, and/or copy their work so long as the authors and source are cited and resulting derivative works are licensed under the same license. Statutory fair use and other rights are in no way affected by the above. Read more about the license at http://www.creativecommons.org/licenses/by-sa/4.0

Cover Art, figures, text and other media included within this book may be under different copyright restrictions. Please see the *Acknowledgements* section at the back of this book for more information.

Cover Image: *The Flamarion Woodcut*, Color by Hugo Heikenwaelder, 1998 CC-BY-SA

PRINT ISBN 978-1-78542-070-2
PDF ISBN 978-1-78542-069-6

OPEN HUMANITIES PRESS

OPEN HUMANITIES PRESS is an international, scholar-led open access publishing collective whose mission is to make leading works of contemporary critical thought freely available worldwide. More at http://openhumanitiespress.org

Contents

1. Introduction: Constructivist Angles   7
2. The Chimera of Relativism: A Tragicomedy   24
3. Religion, Science, and the Humanities: An Interview   38
4. Anthropotheology: Bruno Latour Speaking Religiously   46
5. What Was "Close Reading"?:
   A Century of Method in Literary Studies   66
6. Scientizing the Humanities: Shifts, Collisions, Negotiations   85
7. Perplexing Realities:
   Practicing Relativism in the Anthropocene   106

Notes   125

Works Cited   137

Acknowledgements   149

Chapter One

Introduction: Constructivist Angles

The essays assembled here are about differing accounts of science, belief, and the humanities and, especially, about differing views of their actual, proper, or desirable relations. The decade during which they were written has been a time of enormous technological expansion across the globe, of political, social, and economic upheavals in the West as elsewhere, and of correspondingly significant shifts in the intellectual world. All these have brought painful collisions of aims, styles, and practices in both the sciences and the humanities. These essays—on clashes between established and innovative views of knowledge and cognition, on different ways to frame the relation between "science" and "religion," on new configurations of the sciences and the humanities, and on the divergent realities of climate change—address interconnected aspects of these shifts, strains, and collisions. The title of this volume and, relatedly, that of this chapter say something about the perspectives from which they do so. The terms "relativism" and "constructivism," like all "ism" labels, are only minimally informative and, in both these cases, also subject to recurrent misunderstanding. As I seek to suggest here, however, the views they name, when duly elaborated and understood, can be endorsed without apology, and the angles they offer can be illuminating and productive.

I

In the broadly accepted sense of the term used here, *constructivism* is a way of understanding the relation between what we call "knowledge" or "beliefs" and what we think of and talk about as "reality." In constructivist accounts of that relation, the specific features of what we experience as "the world" or "nature" (objects, events, entity-boundaries, properties, categories, and so forth) are viewed not as independent of our sensory, perceptual, conceptual, and discursive activities but, rather, as built up, specified, and articulated—or, as it is said, "constructed"—through those and other of our activities.[1] Accordingly, the products of cognition, "knowledge" or "beliefs," are seen not as correct-or-incorrect representations of the autonomously existing features of a given world, but, rather, as linked perceptual-behavioral dispositions continuously strengthened, weakened, and reconfigured through our ongoing, more or less effective, interactions with our particular physical and social environments. Also accordingly, what we regard as specifically *scientific* theories or beliefs are seen not as the duly epistemically privileged products of a distinctive set of truth-directed procedures but, rather, as the contingently stable products of especially tightly linked, mutually shaped, conceptual, discursive, and material practices pursued in accord with practically effective methodological traditions. The phrasings here are mine, but these descriptions represent views proposed and developed over the past century by theorists in a range of fields, including the psychology of perception, theoretical biology, philosophy of mind, and the history, philosophy, and sociology of science.[2]

Constructivist epistemology is often identified with "social constructionism." Distinguishing between the two is difficult because both terms have shifting, overlapping usages and the various views named by each have complex intellectual-historical connections. Nevertheless, their simple identification obscures important differences of origin, emphasis, and intellectual operation. In the sense of the term intended here, constructivism has been developed largely by theorists interested in the dynamics of knowledge formation, either at the individual level, as in the cognitive sciences and philosophy of mind, or at the sociohistorical level, as in the history and sociology of science. Social constructionism, as the term is widely used, is a rather differently motivated and differently directed enterprise. As practiced by cultural critics and other scholars in the humanities and social sciences, it operates largely in connection with efforts to expose questionable assumptions associated with prevailing social practices. Thus, when constructivist historians of science speak of the social construction of scientific knowledge, they are commonly emphasizing the complex collective, dynamic, and institutional aspects of the formation of such knowledge, aspects ignored in conventional conceptions of science as an intrinsically truth-directed enterprise pursued by individual agents. When, on the other hand, cultural critics speak of the social construction of racial or gender distinctions, they are commonly emphasizing the culturally mediated forces involved in the perception and maintenance of such distinctions, forces often obliterated or denied in efforts to preserve and justify existing social hierarchies.

Conflations or confusions of constructivist accounts of scientific knowledge with social constructionist cultural criticism helped inflame the "science wars" of the 1990s and continue to be produced and circulated.[3] Although the intellectual clash between constructivist and realist-rationalist epistemologies is genuine, the so-called wars, certainly the idea of widespread "attacks on science" or "abandonments of reason" by trendy academics and French intellectuals, were always something of a mirage. More than two decades have passed since the initial volleys were fired and the early coups have been followed by increasing intellectual irrelevance. Constructivist challenges to traditional views of scientific knowledge remain controversial but are no longer routinely equated with sophomoric slogans and glibly dismissed. The empirical study of science by social scientists—or, as it is now known, science and technology studies (STS)—is an internationally established field producing widely respected research and theory. As younger generations of practitioners, including philosophers, elaborate constructivist views of science, knowledge, cognition, and belief, even the once harrowing charge of relativism has begun to lose its bite.

Suggestions of the contingency, relationality, or contextual dependence of such venerated qualities as truth, scientific validity, or moral virtue have long been seen and represented as "amounting to" or "entailing" a plainly self-refuting position commonly labeled "relativism." Understood as the idea that all views, objects, and practices are equally true, good, proper, and so forth or as the refusal of all value judgments, the position, insofar as it exists, is, of course, foolish and objectionable. The qualification ("insofar as…") is required because, as I have argued for some time and detail here in chapter 2 ("The Chimera of Relativism: A Tragicomedy"), "relativism" thus understood is largely a phantom heresy, continuously generated by the seesaw logic of much orthodox thought itself: if not classic realism, then classic idealism; if not absolute objectivism, then absolute subjectivism; if not one uniquely valid judgment/theory, then all equally valid judgments/theories; and so forth.[4] The phantom may be given added apparent substance, however, by such heavy rhetorical scaffolding as glib conflation, crude decontextualization, tendentious paraphrase, and slapdash intellectual history.[5]

While the foolish ideas and refusals so labeled have few citable, quotable, flesh-and-blood advocates, they are commonly attributed to those advocating quite different, usually philosophically unorthodox but often reasonable and desirable ideas and attitudes. Among the latter are empirically instructed, conceptually well-honed views to the effect that what we take to be good, true, proper, important, and so forth depends on and varies with, among other things, our assumptions, expectations, and categories as these are shaped by, among other things, our inevitably individual experiences and situations in historically and otherwise particular worlds. Accordingly, and as the practice indicated in the title to this volume, "relativism" can be understood as a cultivated consciousness of the irreducible plurality of human perspectives and the contingency of all value judgments. Thus understood, the term names an intellectual disposition in close accord with the constructivist views of knowledge and belief outlined above and elaborated in the chapters that follow.

Along with these views and dispositions, practices of *symmetry* are a recurring theme in this volume. As distinct from more or less dubious claims of equality, they involve acknowledgments and elaborations of significant similarity, correspondence, and/or continuity where sharp distinction and hierarchized difference are otherwise claimed or assumed. Such practices are especially desirable in conjunction with efforts to describe and explain social, behavioral, or historical phenomena, where broadly self-centric (for example, Eurocentric, androcentric, presentist, or "whiggish") accounts are likely to be prevalent. Thus, in fields such as anthropology, historiography, and sociology, practices of symmetry operate as forms of disciplinary self-discipline or, one could say, as efforts at impartiality and objectivity.

An early expression of principled epistemic symmetry appears in the proto-constructivist study, *Genesis and Development of a Scientific Fact*, by Ludwik Fleck (1979 [1935]). Fleck, a microbiologist and medical historian, rejected the idea that there are specific, identifiable features of genuinely scientific knowledge that mark it off clearly from so-called primitive understandings of the world and from earlier specialized knowledges now seen as erroneous. "Whatever is known," he wrote, "has always seemed systematic, proven, applicable, and evident to the knower. Every alien system of knowledge has likewise seemed contradictory, unproven, inapplicable, fanciful, or mystical. May not the time have come to assume a less egocentric, more general point of view?" (23). Fleck's thoroughly historicist and reflexive tracing of the emergence of a scientific fact (his example is the identification of a pathogen for syphilis) is the product of just such a point of view. As such, *Genesis and Development* has figured as a model of constructivist historiography in science studies and, in mainstream epistemology, as an example of "'extreme' epistemic relativism."[6]

The so-called symmetry postulate was initially formulated by Edinburgh sociologist David Bloor as one among a set of four "tenets" (*causality, impartiality, symmetry*, and *reflexivity*) defining a new program in the sociology of scientific knowledge (Bloor 1976). In contrast to rationalist philosophy of science and prevailing sociologies of science, both of which saw "social" explanations as relevant only to erroneous theories or to the resistance to correct ones, the "Strong Programme" would investigate the social forces involved in the fortunes of all scientific theories, correct as well as erroneous, true as well as false, rational as well as irrational, including, in principle, the theories it produced itself. Misstated as the claim that all theories are equally true or that scientific theories are mere reflections of social forces, the symmetry postulate has been recurrently cast as a naive and debilitating relativism and dismissed accordingly. Thus philosopher Paul Boghossian, commenting on Bloor's views and constructivist sociology more generally, writes, "[I]f you say that the correct explanation for all scientific belief is in terms of the political goals, interests and prejudices of the scientist, you make it impossible to criticize a specific scientific belief as merely reflective of such prejudice" (Boghossian 2002, 223). Of course, neither Bloor nor any other constructivist sociologist says that "the *correct* explanation for *all* scientific belief is in terms of the *political* goals, interests and prejudices of *the scientist*." Nor do constructivist accounts of the formation and stabilization of scientific knowledge make relevant discriminations and challenges impossible. Rather, by making visible the complexity of the processes through which scientific (and other) beliefs are communally established, including the multiplicity of agents and interests involved and the significance of specific social, institutional, cultural, and discursive conditions, they illuminate controversies over the validity of specific knowledge claims, present as well as past, and indicate the types of activity required either to establish them communally or to challenge them effectively.[7]

The symmetry postulate and other tenets of the Strong Programme were not initially well phrased or well explained, and, as the program was pursued by researchers and theorists in the United Kingdom, the United States, and Europe, important clarifications, elaborations, and modifications followed.[8] In a series of radical modifications that came to be called actor-network theory (ANT), French philosopher/sociologist Bruno Latour advocated a "generalized" symmetry principle in accord with which the consequential activities and operations of objects and nonhuman animals would be treated symmetrically with those of human agents in accounts of the emergence of scientific facts: for example, sick cows and petri dishes acknowledged along with scientists and physicians in the emergence of the microbe theory of disease. Understood as promoting the moral status of animals as equal to that of humans and as granting due recognition to the previously ignored or slighted capacities of "things," the generalized symmetry principle contributes to current embraces of Latour as a posthumanist "object-oriented" thinker and to appropriations of ANT as a properly non-anthropocentric program of research. However dubious some of these understandings and appropriations,[9] Latour's engagements with ideas and practices of symmetry have been important in the development of constructivist science studies, and his more recent metaphysical ventures and concern with environmental issues have figured significantly on the intellectually scene more broadly. His treatments of scientific knowledge and religious belief are examined here in chapter 4, and his influential responses to the epistemic and rhetorical challenges of climate science are discussed in chapter 7.

II

An interest in intellectual controversy and an emphasis on the radical contingency of value judgments have both been central to my work for some time, partly in response to the "canon wars" in the Anglo-American literary academy but more broadly, in *Contingencies of Value* (Smith 1988), as an effort to develop conceptually workable alternatives to dubious ideas of intrinsic, objective, or universal value. A subsequent book, *Belief and Resistance* (Smith 1997), deals with the complex social and psychological dynamics of belief and skepticism in current epistemological controversies, and I extend its major observations to debates over religious beliefs in my 2006 Terry Lectures, *Natural Reflections: Human Cognition at the Nexus of Science and Religion*. Publication of the latter (Smith 2009) was the occasion for the interview in chapter 3 ("Religion, Science, and the Humanities").

Natural Reflections examines two contemporary intellectual ventures involving science and religious belief. One is a set of efforts by anthropologists and scholars of religion to "explain" religion, as it is said, "scientifically," here largely on the basis of claims about human behavior emerging from evolutionary psychology. I call it the New Naturalism. The other is a set of efforts by scientifically informed theologians to demonstrate the compatibility of scientific knowledge with traditional religious teachings, here through theistic readings of selected concepts in the natural sciences. I call it the New Natural Theology.[10] Practitioners of both ventures, I suggest, exhibit certain endemic cognitive tendencies, including confirmation bias, dissonance avoidance, and dualistic thinking. As noted by the interviewer Nathan Schneider, a report of the book in *The New York Times* (Fish 2010) attracted considerable attention from readers. Based on the report, many of them took my evenhanded treatment of the explanation-claiming social scientists and compatibilist theologians as endorsing one or the other side of a supposed fundamental conflict between "Science" and "Religion," both conceived in simplistic, monolithic terms. Several others took my notice of the coexistence of scientific and religious practices in the lives of many people as endorsing biologist Stephen J. Gould's notion of science and religion as "non-overlapping magisteria" (Gould 1999). Commenting on these misunderstandings in a subsequent column (Smith 2010), I note that while, in Gould's account, science and religion remain monoliths or, as one clever reader puts it, "rocks of ages" (Gould's book is itself titled *Rocks of Ages*), my own efforts—in effect, to pulverize both rocks—are quite different.

The framings of the relations between science and religion in Bruno Latour's writings are strongly, indeed strikingly, symmetrical. As I indicate in chapter 4 ("Anthropotheology: Bruno Latour Speaking Religiously"), Latour gives uncommon care and imagination to elaborating their parallels and correspondences. Also, here in company with, among others, Paul Feyerabend (1975) and Paul Veyne (1988), he stresses both the comparable constructedness of the entities of each (scientific facts as well as religious fetishes) and the comparable contingency of their existence (angels as wells as quarks). Fairly uniquely among theorists of science, however, Latour has sought to make religious experience vivid for his audiences. And, in a major work (Latour 2013), he insists, questionably in my view, on both the distinctness of the "mode of existence" of the entities of each and the incommensurability of their respective "modes of veridiction." Latour explicitly abjures the word "belief" (*croyance*) here.[11] Acknowledging the difficulties of such a drastic refusal and possible objections to it, he explains the move as follows:

> *Belief* is obviously multimodal and many of its significations are innocent, but the word also designates the attribution of a mental state to those we encountered during colonial exploration. This state of mind ... [depends] on an epistemological division imposed a priori between truth and illusion. It is this division which justifies a desire to put an end to beliefs and to destroy illusions, idols, in the name of anti-fetishism.[12]

The reference here is to colonialism and anti-fetishism, but Latour elsewhere indicts more general devaluations of religion—and the iconoclasms that attend them—issued in the name of enlightenment, rationality, and science. These devaluations commonly involve the drawing of strong contrasts between supposedly distinct mental states or processes, and the contrasts are commonly aligned with sharp distinctions between what are specifically named scientific "knowledge" and religious "belief."[13]

One can sympathize with Latour's desire to bar such contrasts. As he suggests, they draw on views of mental life that are mired in dubious rationalist assumptions and perpetuate simplistic notions of scientific thought as well as of religious experience. Many of the views in question are ancient but transported into contemporary discourse by rationalist traditions in philosophy of mind and related empirical disciplines. Important challenges to those views have been mounted over the past century by historians and sociologists of science—Latour among them— and by some dissident philosophers of science. As indicated above, however, related challenges have been mounted on other fronts, including the psychology of perception, theoretical biology, and some dissident regions of cognitive science and philosophy of mind. Latour rejects naturalizing explanations of religion and writes scornfully of efforts to explain subjectivity in terms of "genes" or "neurons." Research and theory in the fields just mentioned, however, are worth his attention and that of the science-studies community more broadly. The challenges they raise to rationalist assumptions are powerful, and their alternative accounts of cognition have important implications for how we understand subjective experience, knowledge, and belief.

Mainstream cognitive science, shadowed by classical epistemology, seeks to explain how, as it is said, we *acquire accurate knowledge of the world*, the crucial assumption being that such knowledge is required to navigate the world and that humans, perhaps uniquely, are equipped in some way to obtain it. The question might be less tendentiously and more usefully phrased, however, as how do we, like other organisms, *come to operate effectively in our worlds*. Phrased this way, cognition is understood not as the production of more or less accurate internal representations of a fixed, given, exterior world but as an ongoing, dynamic process involving the entire organism interacting with its specific environment. This pragmatist-constructivist understanding of cognition has multiple names and important variants, among them, "ecological," "embodied," "dynamic," and "enactive."[14] In virtually all its variants, however, it has been developed as an alternative to traditional rationalist views.

The understandings of cognition referenced above offer compelling ways to reframe the familiar contrasts drawn between scientific and religious ideas: "genuine knowledge" versus "mere belief," "reason" versus "irrationality," and so forth. Among other things, they suggest that religious ideas are formed not through any special cognitive process, either especially defective as rationalists suggest or uniquely transcendent as theologians may claim, but through the same general processes as any of our ideas or beliefs, from everyday notions to scientific or philosophical theories.[15] Though distinctive in many ways, the various entities constructed through those processes, from the scientist's quarks and molecules to the philosopher's rocks and tables and the religionist's demons or divinities, would therefore be equally open to symmetrical, commensurable explanation. These views of cognitive process also suggest that the dispositions and activities associated with doing science and with practicing a faith are continuous with each other and with the dispositions and competencies exhibited in other cultural, social, or personal domains, from art and music to commerce and politics. Contrary, then, to familiar theological claims and rationalist assumptions, what are commonly called "religious" beliefs and experiences are not unique, either ontologically or cognitively, and their elaborations and associated practices are not altogether distinct from those found elsewhere in our cultures and lives.[16]

In a lengthy entry on "psychology" in the online version of *An Inquiry into Modes of Existence* (AIME), Latour characterizes the field as the improperly "interiorizing" counterpart to what is now widely recognized as the improper "exteriorizing" of epistemology.[17] The suggestion is that, while classical epistemology explained the manifest success of scientific knowledge by positing the lawful operations of a supposedly objective Nature, psychology explains the presumed irrationality of religious belief by positing the wayward operations of a subjective Mind. In some respects, Latour's dismissal of psychology recalls the rejection of "psychologism" in early twentieth-century phenomenology, where the argument was that the objects of phenomenological investigation could not be explained using the methods or appealing to the empirical concepts of the experimental psychology of the time (Kusch 1995). In this regard, it also recalls two other perplexities of contemporary interdisciplinary and cross-cultural research. One is the "hard problem" of contemporary philosophy of mind: that is, the challenge of explaining in observable, physical terms what is experienced in consciousness. The other is the *emic/etic* distinction of ethnography: that is, with regard to indigenous practices, the fundamentally different perspectives of the native participant and the observing anthropologist. At least part of the difficulty in each of these cases is translating not merely between different verbal and conceptual idioms, which can be "hard" enough, but also between deeply different and sometimes mutually incompatible conceptual tastes and ontological commitments.

Such cross-disciplinary and cross-cultural translations require a good bit of intellectual and ethical tact. In relation to the new forms of interdisciplinarity between the sciences and the humanities, such tact would involve not merely recognizing different disciplinary aims and perspectives but also acknowledging crucially different intellectual histories and personal temperaments among practitioners and being symmetrical about all these in relation to one's own. And, in regard to the sometimes dubious cosmopolitics proposed for the Anthropocene, it would involve not the recognition of *common* interests or attachments but the search for practically *congruent* ones, a goal that seems at least conceivably achievable.

III

A mood of distress is evident across the humanities, along with efforts to update and transform those disciplines—and, sometimes, thereby to redeem them—via closer connections to the sciences. In some respects, these developments involve head-spinning reversals. Where, in the past, one heard claims of the creative, conceptual, and moral superiority of the humanities to the merely mechanical, merely numerical, merely utilitarian natural sciences, one now hears assertions of the epistemic superiority of science to the merely frivolous, objectionably elitist, manifestly obsolete aims and methods of the humanities. Or where, in the past, the biological sciences were understood to be fundamentally complicit with sexist, racist, imperialist ideologies, Darwin's writings are now closely studied as, among other things, especially illuminating for feminism. The new ecumenical attitudes toward the natural sciences on the part of humanities scholars are, in my view, altogether to be welcomed. But two-culture caricatures remain in play across the academy as well as in the culture at large, and they contribute to a new or newly energized scientism now displayed *within* the humanities as well as *against* them. The epistemic aims and claims in play in some of these programs, and especially in their promotion, invite critical attention. I consider them in chapters 5 ("What Was 'Close Reading'?: A Century of Method in Literary Studies") and 6 ("Scientizing the Humanities: Shifts, Collisions, Negotiations").

The oversimplifications I noted above in clashes over the knowledge-claims of science and religion recur in clashes over how the sciences are related to the humanities, with some additional twists. Distortions in both cases reflect a long history of polarized opposition, with ongoing disputes over which should be epistemically privileged. Both pairings require, accordingly, due attention to the multiplicity and heterogeneity of the practices designated by each term and also to the complexity of their respective historical, institutional, and cultural connections. Thus "science," as in the title of the promotional book *What Science Offers the Humanities* (Slingerland 2008), should be pluralized in these debates, not only to mark the parallel with the humanities but also to acknowledge just that *double* multiplicity and heterogeneity.[18] In view of the specific obliterations attending current promotions of the integration of the humanities with the sciences, it is also important to recall their historically different aims and functions. Attempts to articulate those differences face considerable conceptual and rhetorical difficulties, however, at least if they seek to be duly symmetrical.

Many of the arguments promoting new methods and approaches in the humanities cite huge intellectual changes or "tectonic" shifts, mostly scientific and technological advances, as requiring the proposed transformations. But just about everything else in the academy has been changing as well, with consequences not always recognized either by those promoting radical reforms or by conservative defenders of traditional methods and missions. The various explicit missions of literary study in the Anglo-American academy—aesthetic appreciation, moral edification, cultural education, and so forth—have always depended on assumptions about the broader social functions of such study, especially in relation to classroom teaching, the de facto area of the "application" of literary research. The pursuit of those missions has depended, accordingly, on what faculty knew or believed about the capacities and interests of students, which could, and regularly did, become out of phase as both student and faculty demographics changed. Thus, as I note in chapter 5 in connection with the presumed aims of literary study, preparing the sons of the upper classes for leadership in the professions in the early twentieth century was different from providing upward social mobility for the sons and daughters of immigrants in the 1930s or job credentials for war veterans in the postwar years.

Changes outside the academy have also always affected understandings of the role of the humanities and their value in the larger society. Important changes in recent years include immigrations from increasingly diverse regions of the globe and, with them, increasingly complex cultural encounters; radically new art forms and media and, with them, radically different ways of becoming acculturated; newly conspicuous global phenomena, environmental as well as political, and, with them, intellectual and ethical perplexities on a scale not previously experienced. It may be defensible but it is not obvious, under these conditions, that humanities education in North America should focus on the canonical works of the Western literary, philosophic, and artistic tradition.

Critical, pedagogic, and scholarly practices in the humanities also change in response to more or less independent intellectual and cultural developments, and faculty recurrently find their own talents, training, and past achievements out of phase with currently dominant interests and approaches. In the 1980s and 1990s, the appearance of deconstruction, cultural studies, postcolonial criticism, and queer theory made professional life difficult for those in literary studies with standard postwar training. Never mind the historical missions of moral edification and aesthetic cultivation: what was one to do when one's specialty was Golden Age Spanish literature and all one's colleagues and students seemed to care about was Chicano film studies? Not surprisingly, professional life can also become difficult when one's specialty is Chicano film and all one's colleagues and students seem to care about is digital games and posthumanist animal studies.

Conservative defenses of traditional practices in the humanities often charge the new affiliations with the sciences with "scientism," sometimes, but not always, with good reason. Since the term, like other "isms" discussed above, is subject to variable usage, it requires some attention here. An appreciative engagement with concepts and findings in the natural sciences in studies of art, literature, or religion does not, in my view, warrant the label "scientism" or make such studies dismissible on just that count. On the contrary, the encouragement of such interdisciplinary engagement has been a focus of much of my professional life. What I characterize as such in these essays as elsewhere is the idea or evident assumption that the aims and methods of the natural sciences should be taken as models for all knowledge practices. The idea was articulated influentially by biologist Edward O. Wilson in his 1998 book *Consilience: The Unity of Knowledge* and is often attended, as in Wilson's book, with the idea that the humanities disciplines are at best prescientific and should be shepherded as quickly as possible, along with some still vagrant social sciences, into the fold of the natural sciences. To those who view science this way (not only or always scientists) and who understand the humanities as defined only by the objects they study, this evidently makes perfect sense: people who study artworks, literary texts, historical records or anything else should be studying them scientifically. But if one views *the sciences* in the ways suggested by a century of research and theory in constructivist history, sociology, and philosophy of science and conceives *knowledge* in the ways suggested by a century of research and theory in cognitive science and constructivist epistemology, then one might be prepared to see the humanities disciplines as defined not by the objects they study but, as I detail in chapter 6, by their distinctive epistemic orientation, in which case their absorption by the natural sciences makes no sense at all.

IV

A number of themes outlined above—rival epistemologies, the complex dynamics of belief and skepticism, practices of symmetry—figure centrally in chapter 7 ("Perplexing Realities: Practicing Relativism in the Anthropocene"). Written initially for a conference on Climate Realism, it argues that being realistic about climate change does not require an endorsement of philosophical realism or an abandonment of critical practices with regard to the natural sciences or anything else. On the contrary, the views of reality and accounts of science likely to be most serviceable in these connections are to be found where philosophical realism has been most effectively challenged: that is, in science and technology studies and in the work of constructivist theorists—academic philosophers and others—engaged with developments in related empirical fields. Accounts of these matters offered in the mainstream regions of such fields (cognitive psychology, neuroscience, behavioral ecology, and so forth) continue to be hobbled by dubious assumptions and concepts. That is not good reason, however, for humanities scholars to dismiss the accounts or the fields. On the contrary, it is good reason for them to engage work in those fields, precisely, *critically*: to examine the accounts carefully, to invoke the findings with due discrimination, and to contribute to the ongoing reformation of the dubious assumptions, concepts, methods, and explanations. In relation to the current ecological crisis, along with economic crises, ethnic conflicts, the "dark sides" of information technologies, and other matters of broad contemporary concern, it is likely that work in such empirical fields will be more conceptually useful and potentially politically transformative than an inevitably limited number of metaphysical or spiritual conversions, however philosophically sophisticated their inspiration.

With regard to denials of global warming, environmentalists have come to recognize not only the significant operation of powerful economic and material interests but also the existence of deeply vested cognitive interests and the power of quite primitive endemic impulses, cognitive and other. Not all the failures of acknowledgment of climate change on the part of the general public are the product of false beliefs deliberately promoted by energy moguls. Other social and psychological forces, more elusive but no less powerful, are clearly in play here as elsewhere. Indeed, the efforts of several environmental activists are now directed at detailing such forces, often in the hope that widespread knowledge of them will help counter their general operation. Though I am wary of talk of human universals, I have entertained for some time the idea of two general tendencies that seem to explain a good deal of human history. The first is *People would rather continue believing what they've always believed than believe something different*. The second is *People would rather think well of themselves than otherwise*. Clearly the first tendency is involved in climate denialism. In view, however, of current political directions in the United States and elsewhere, there is reason to think that the second may matter at least as much. Irrational behavior often reflects an effort to escape the cognitive dissonance of radically new news. But it can also reflect the rage of resentment, the conviction—justified or not—of insult beyond injury. If there is a lesson here for environmentalists, it is that formidable resistance to enlightenment about climate change may be found not only among the uninformed and misinformed but also among losers of place or face.

One of the major perplexities created by the complex state of affairs that we, or some of us, name climate change is the inevitable multiplicity of *operative* realities: not many ways of perceiving a shared world but many worlds in the sense of irreducibly divergent experiences, percepts, constructs, and cosmologies. In influential writings, Bruno Latour has suggested that the way to arrive at the commonality he sees necessary to address the realities of the Anthropocene is by overcoming existing conceptual divisions—drawn, in his view, by Enlightenment metaphysics—between subjects and objects, humans and nonhumans, and nature and society. To accomplish this, he writes, "we should abstain from de-animating the agencies that we encounter at each step" (Latour 2014, 14). Thus and only thus, he maintains, shall we be able "to move out of the impasse in which modernism has dug itself so deeply" (15). No doubt, with regard to the ongoing degradation of the environment and the threat of planetary ecological catastrophe, the conceptual divisions that concern Latour contribute to the thinking involved in the current political paralysis and perpetuation of disastrous practices. As I suggest above, however, there are other sources of collective inertia and self-destruction that are at least as relevant and perhaps even more powerful. The cognitive liabilities and other primitive tendencies noted here are not confined to Moderns. Nor can they be said to stem from Enlightenment metaphysics or from what Latour, among others, indicts as modern thought, or, for that matter, from present political arrangements as argued by other theorists of the Anthropocene. Moreover, the tendencies involved, both typically and historically, are not lamentable flaws of human character. They are fundamentally double-valued human dispositions, individually and collectively advantageous under many conditions, disadvantageous under many others. There is no point to denouncing them. Nor, I think, should we hope or expect to eradicate them, whether through species-wide self-transcendence, efforts at a counter-enlightenment, or otherwise. We might, however, strive to be aware of their powerful operation in ourselves as well as others so as, perhaps, to be more disposed to act, individually and in concert, in better ways rather than worse.

Angles are not solutions. Angles are perspectives. Solutions to problems of the kind evoked here can only come out of ongoing engagements with specific conditions and can only be provisional ways of opening out—not, in my view, to utterly new ways of being but to relatively new ways of thinking and to better ways of continuing. What this volume suggests is that those ways are likely to be better for all concerned, nonhumans included, when, as the angles offered here may incline us, we are especially alert to the contingency, complexity, and multiplicity of our worlds and to our own all-too-human ways of being in them.

Chapter Two

The Chimera of Relativism: A Tragicomedy

Whatever sort of thing relativism is taken to be—doctrine, thesis, crime or folly, insight or abyss—it is certainly, from the perspective of intellectual history, exceptionally elusive. Is there a single specifiable claim or denial, even a minimally describable "family" of them, shared by Protagoras, Montaigne, Nietzsche, Heidegger, Franz Boas, Paul Feyerabend, Jacques Derrida, Bruno Latour, and the majority of undergraduates on today's college campuses? All these have been said to "embrace" or "espouse" relativism, or to have "slipped" or "fallen" into it, by reason of some utterance they have made or failed to make, some attitude they have displayed or failed to display.[19]

Logicians suggest that we are in the presence of relativism when one or another self-evidently solid and important thing (for example, truth, value, meaning, or reality) has been said by some perverse or logically injudicious person to be "relative to" something soft or slippery (for example, context, culture, language, conceptual scheme, individual perspective, or political interest). But such X- is- relative- to-Y statements are almost always distortions of what has actually been said. What has often actually been said—as in the case of the figures I have mentioned, from Protagoras to Latour—are statements to the effect that human perceptions, interpretations, and judgments are not absolute, universal, or objective in the sense of being independent of all perspectives and/or invariant under all conditions; that what we take to be real, true, and good depends upon and varies with, among other things, our assumptions, expectations, categories, and existing beliefs as these are affected to one degree or another by, among other things, our particular experiences and situations, both past and ongoing; and that these in turn are affected to one degree or another by, among other things, our historically and otherwise particular social, cultural, and institutional environments, including the conceptual and verbal idioms prevailing in our communities.

These statements of variability and contingency have challenging implications for familiar ways of talking and thinking about truth, value, and reality; but they are not claims about any of these terms or concepts taken as autonomous entities or properties. Statements about the dependence of our perceptions, interpretations, and judgments on various more or less unpredictable and uncontrollable sets of conditions do not amount to denials of the existence of anything that could be called truth or to claims about the purely subjective, purely verbal, or purely social status of things such as stones, mountains, quarks, germs, or genes that we may take to be, in some unproblematic sense, real. Rather, statements of these kinds, often said to be relativistic, alert us to the relational aspects of seemingly autonomous entities and seemingly inherent properties and to the fact that the quite heterogeneous situations that we name *truth, fact, knowledge, science,* or *reality* are often quite complexly constituted and sustained. Such statements also alert us to the historicity and the often far-from-unproblematic meaning of such ideas as objective truth or transcendent value and to the deeply problematic nature of reality if it is understood as an autonomous, absolutely privileged realm of being.

The sorts of statements just described are not shallowly reductive, deterministic, or gloomy. They do not claim that human perspectives vary only historically or culturally; they do not require us to believe that the human mind is a blank slate; they do not commit us to the view that people from different eras or cultures have nothing in common, or that each culture or period is a distinct and isolated universe, or that the differences among our perceptions, interpretations, or judgments are always very large or very significant. I note all these negatives because the sorts of observation that I am describing here—observations of contingency, variability, and/or relationality, often labeled relativist or taken as reflections of an implicit doctrine of relativism—are commonly paraphrased in just these ways or said to have just such dubious or unhappy implications.

I

In the human sciences, what often elicits the charge or label "relativism" is the display of certain attitudes, notably those of epistemic tolerance, or the recommendation of certain methodological principles—especially, in recent years, principles involving explanatory or evaluative symmetry. The two disciplines that have proved most generative of the attitudes and principles in question are history and anthropology or, more precisely, historiography and ethnography—the charting of other times and other people. Of course, the observation of human variety in travel, including time travel, does not always increase tolerance or chasten egotism. It can just as readily deepen misanthropy or ratify a sense of the perfect reasonableness and absolute propriety of one's own views and practices. Nevertheless, reports of what seem to be other ways of being human have operated virtually from their beginning as a reservoir of counterexamples to standard views of what is natural, necessary, or inevitable for members of the species to do, feel, or think.

Along with more particular forms of sophistication arising from their work in the archives and in the field, historians and anthropologists often develop a generally heightened consciousness of the variability of human practices, institutions, and individual responses. While a sharpened awareness of this kind may develop from everyday observation, as reflected in the prudential axioms of folk relativism (*different strokes for different folks* and so forth), of particular interest here is the explicit cultivation of such awareness in the pursuit of disciplinary aims as recommended by such influential early twentieth-century historians, anthropologists, and social theorists as Émile Durkheim, Max Weber, Carl Becker, and Ludwik Fleck and, later in the century, by such important figures as Benjamin Lee Whorf, Thomas Kuhn, and David Bloor.[20] While not all these figures described their views or programs as relativistic and none proclaimed any doctrine under the label of "relativism," each stressed the dependence of our ideas and practices on historically and culturally variable conditions and on individual perspectives, and all emphasized the consequent need to cultivate a self-conscious wariness in their respective fields. These scholars and theorists made such points against what they saw as dubiously self-privileging aims, claims, and methods in their particular fields, and their remarks were directed more or less exclusively and explicitly to fellow practitioners. The principles in question were not, in other words, produced out of the blue, nor were they directed to humanity at large or intended to govern the general conduct of ethical or intellectual life.[21] This last point requires special emphasis because it is routinely missed by commentators who, after plucking those recommended principles from their intellectual, institutional, and historical contexts and improperly absolutizing and universalizing them, go on to register outrage or alarm at the absurd, unwholesome, or debilitating implications that they thereupon derive from them—the implication, for example, that all beliefs and cultural practices are equally worthy of respect or that "reasons" cannot or should not count in our assessment of truth claims. Indeed, the obliteration of relevant historical, intellectual, and institutional contexts is crucial in the generation of the doctrines, "claims," or "theses" that make up the chimerical beast—part straw man, part red herring—commonly evoked under the name "relativism."

This chimera—or, as I have called it elsewhere, fantasy heresy—is not the product of dishonest or intellectually incompetent people. On the contrary, it is largely the issue of critical efforts by intelligent, sometimes ethically motivated, often exceptionally well-trained people. What must be added, however, is that they are trained in quite particular conceptual idioms. Such idioms may be well established in their own fields and serve their professional purposes satisfactorily in the domains of their customary disciplinary practices. Nevertheless, the concepts that are central to those idioms, and the conceptual syntax that is central to their deployment in descriptions, analyses, and arguments, are themselves historically contingent and local, not necessary or universal.

An example of the significant operation of such conceptual idioms can be seen in an argument offered by philosopher John Searle, reviewing and endorsing philosopher Paul Boghossian's book, *Fear of Knowledge: Against Relativism and Constructivism* (2006). Hypothesizing what a Native American might say about how his tribe originally came to occupy its lands, Searle argues that it is a "requirement of rationality" that "anyone who makes such a statement is thereby committed to the existence of a fact," that this "commitment in turn carries a commitment to being able to answer such questions as ... What is the evidence?", and that "only certain kinds of things can count as evidence" (Searle 2009, 91-92). For Searle, as for Boghossian, it is self-evident that the things that can count as evidence do not include traditional accounts transmitted in tribal legends. Accordingly, neither of them can understand—and both therefore regard as merely ideologically motivated—a claim to the effect that traditional accounts can and do count as evidence for members of the tribe in the sense of having conviction-affecting weight for them. Of course, tribal legends do not count as evidence in United States courts, or in the arguments of most academic philosophers, or when anthropologists address each other regarding scholarly claims. But the reasons they have no evidentiary weight in these contexts have nothing to do with something "built into the fundamental structure of thought and language." Nor do they have anything to do with "the requirement[s] of rationality" except insofar as *rationality* is itself understood in historically and otherwise quite particular ways: here, in ways that operate with inter-validating reference to terms like *statement, fact,* and *evidence*.[22]

Relativism-refuting scholars typically take for granted certain definitions, distinctions, and conceptual relations that other scholars—precisely those whom they charge with embracing an absurd or appalling relativism—have come to view in crucially different ways. Because the relativism-refuters appeal in their arguments to those presupposed definitions, distinctions, and relations, their arguments may appear, at least to those who share their conceptual idioms and syntax, to demonstrate that the alleged "claims" of the alleged "relativists" are indeed absurd or appalling. But, for the same reason—that is, because the relativism-refuters deploy and depend on the very concepts and relations that are at issue (concepts such as *truth* and *reason*, relations such as those between what are referred to as *facts* and *evidence*)—their arguments can have no intellectual force for the alleged relativists, who know themselves not to be saying the foolish things they are charged with saying and who do not and cannot take for granted the concepts, definitions, distinctions, and relations to which the relativism-refuters appeal. The result is pure nonengagement and perfect deadlock. The chimerical beast called Relativism lies always already slain but, as one of "the philosophical undead" (Rouse 2002), it always walks again.[23]

There are signs that this tragicomic episode of intellectual history may have run its course. Most cultural anthropologists and historians of science abide by the now well-established methodological principles of their respective disciplines without giving them much thought or finding it necessary to explain or defend them to anyone.[24] At the same time, the scope, force, and interest of formal exposures and refutations of what is named "relativism" appear to be diminished in the current philosophical literature and elsewhere. Indeed, some younger logicians and philosophers now unapologetically detail and defend relativity-affirming theses as such, while, for better or worse, a genial folk-relativist ethos (*live and let live*) seems to prevail among students on many of our de facto multicultural campuses.[25] Where full-throated formal denunciations of relativism continue to be voiced in contemporary discourse, they seem to issue primarily from protectors or would-be restorers of some threatened or faded orthodoxy—for example, positivist scientism, rationalist epistemology, or Vatican infallibilism—and are clearly designed to discredit one or another currently significant challenge.[26]

In spite of these hopeful developments (hopeful, at least, from some perspectives), anxieties about the implications or consequences of what is identified as relativism linger, and there are also recent efforts to refute relativism on what are said to be scientific grounds. Two contemporary sites of antirelativist energy require special attention. The first centers on the claim that cultural relativism is refuted by the demonstrated existence of cognitive universals. The second involves the fear or charge that relativist convictions lead to politically debilitating neutrality in the face of oppression and other social ills.

II

According to an influential group of evolutionary psychologists and cognitive anthropologists, we now have evidence of what they refer to as "the psychic unity of mankind"—specifically, the existence of innate, evolved, universal mental mechanisms underlying all human thought, behavior, and culture (Tooby and Cosmides 1992). Questions can be raised about the empirical basis and conceptual coherence of this view, which is by no means accepted by all psychologists or cognitive scientists. What concerns me here, however, is the attendant argument that the demonstrable existence of such universal mechanisms undercuts certain alleged relativist claims.

The argument just described appears in a book, *In Gods We Trust* (2002), by cognitive anthropologist Scott Atran. Atran observes that "there is a long-standing claim on the 'relativist' side of anthropology, psychology, and the philosophy and history of science to the effect that people who live in 'traditional' cultures ... live in conceptual worlds that are profoundly and incommensurably different from our own world (and each other's worlds)" (84). According to Atran, "this claim is mistaken in light of the following *facts*" (emphasis added): "1. There is considerable recurrence of symbolic content [of supernatural beliefs] across historically isolated cultures ... 2. This recurrence owes chiefly to universal cognitive mechanisms that process cultural input (information) in ways that are variously triggered but subsequently unaffected by the nature of the input." Atran then lists some beliefs recorded among the Itza' Mayans of Mexico that "we" would find hard to believe or to restate in any way that made sense—for example, that a certain sorcerer transformed himself into a dog, that a person "ensouls" a house, and that a house has a soul.[27] On such beliefs, he comments as follows:

> From the forgoing we might conclude that we and the Itza' just live in conceptually different everyday worlds. That people abide such apparently different worlds may, in turn, be taken as support for the flexibility of the human mind, that is, a mind unconstrained by cognitive structures that are evolved ... task-specific or innately determined and content-constraining. But this conclusion is wrong. (86)

The somewhat awkward phrasing here coincides with the tenuousness and circularity of the argument. Atran speaks of a conclusion wrongly drawn from ethnographic data about the strange beliefs of other people and an idea wrongly drawing support from that data. The conclusion is that people can live in conceptually different worlds; the idea is that the human mind is flexible. But is either of these wrong, and are they wrongly concluded? Is it not the case that people, even some who eat daily in the same faculty-club dining rooms, can live in conceptual worlds that are profoundly different—for example, as I suggest above, relativizing anthropologists and relativism-refuting philosophers? And is it wrong to conclude from, among other things, the wide variety of cosmologies encountered by ethnographers that "the human mind"—which of course names a range of capacities and activities—is flexible?

Two other questions are significant for the logical and rhetorical force of Atran's argument. One is whether the general observation that "the human mind is flexible" is properly glossed as the technically particular (and manifestly self-contradictory as stated) claim that the mind is "unconstrained by cognitive structures that are evolved [,] ... task-specific or innately determined and content-constraining" (86). The second is what makes a "fact" out of Atran's explanation for the cross-cultural recurrence of the thematic content of various supernatural beliefs. According to his argument, such recurrences are evidence for the existence of universal cognitive mechanisms, but the only basis he offers for the existence of such mechanisms is the questionable contention that their operation is the only thing that can explain such recurrences. The cognitive-universalist claim and the supposed ethnographic evidence for it are bootstrapped onto each other and the purported refutation is totally circular.

Contrary to Atran's contention, the recurrence of various mythic and religious themes can be explained without positing highly specific, content-constraining cognitive mechanisms. One notes, for example, the existence of such widespread—indeed pan-cultural—phenomena as sunrise and nightfall and the prevalence of such salient objects, events, and experiences as birds and snakes, journeys and warfare, illness and dreams. Also, as described in other chapters here, there are important alternatives to the strongly innatist, adaptationist view of mind and cognition that Atran invokes here as factually established.[28] Such alternative views do not, as he suggests, come down to the blank slate of classic empiricism or to cultural-environmental determinism. They certainly do not claim that the human mind is altogether unconstrained. What they maintain or indicate instead is that, although various species-wide cognitive capacities, traits, or tendencies may exist, they must, in their actual operations, interact continuously with other more or less highly individuated traits and tendencies and also with the traces of individual experiences in particular physical, social, and cultural worlds. The point is significant for the operation of any putative human or cognitive universal, from language ability to the often popularly posited "moral sense."

Theorists and scientists proposing these alternative views of human cognitive or psychological development would reject Atran's reductive and otherwise dubious account of the generation of supernatural concepts by highly specialized mental mechanisms. They would also reject the mind-as-blank-slate view that Atran refers to, somewhat oddly, as "relativism" and that he implies is the only major alternative to his own strongly mentalist account of cognition. The individual "mind," however we understand the term, is shaped by multiple forces, and its operations and products are of course limited or "constrained" in many respects; and "the human mind," taken either individually or as a species characteristic, is flexible. These are not mutually exclusive observations. To maintain that the mind is flexible, in the sense of being responsive and capable of ongoing modification, is not to deny the existence of constraining forces on cognitive processes and products, including forces arising from general features of human neurophysiology as shaped over evolutionary time.[29] The conflict of views that Atran evokes here is spurious. If there are any pure cultural determinists remaining in anthropology, or any strict environmental determinists among behaviorists or Jesuits, none of them has much authority in the current intellectual world.[30] There are real conflicts here, but they are not over whether Nature or Nurture, evolved neurophysiology or culturally contingent experience, is decisive in shaping the content of our beliefs. The conflicts are over the contested institutional dominance of different factions in the contemporary social sciences, with so-called cognitive approaches seeking not only to displace older and no doubt limited approaches but also to hold their ground against newer and arguably more fertile developments.[31] "Relativism" here, as often elsewhere, is a straw herring.

III

I turn now to the other current objection to relativism that I mentioned earlier. This is the fear or charge that relativistic convictions lead to ethically reprehensible neutrality or political passivity. One can understand how the charge or anxiety originates. Where the theoretically described variability of human perceptions, interpretations, and judgments is manifested as consequential conflicts and those conflicts come close to home, efforts by historians, anthropologists, and other scholars to maintain methodological symmetry and to treat all sides evenhandedly may appear improper and will be, from certain perspectives, objectionable. Moreover, efforts at neutrality under such conditions are likely to become strained for the scholars themselves. For example (quite close to home for Americans), an otherwise conscientiously impartial sociologist of science may find it hard to treat current, local promotions of biblical creationism symmetrically with efforts by biology teachers to present evolution without disclaimers in public high schools. Similarly, a Western-educated anthropologist may find it hard to report impartially on such exotic practices as female genital cutting. Under such conditions, the determinedly impartial sociologist of science or symmetry-maintaining anthropologist may be too involved in the outcome of such struggles or too conscious of the effects of such practices on the lives of people he or she knows well, or can imagine vividly enough, to maintain an otherwise proper neutrality.

Where a conflict of views—perceptions, interpretations, or judgments—is neither hypothetical nor in the remote past, but actual, sharp, current, and caught up in one's personal history or sense of personal identity, it is understandably difficult to be impartial. Where, moreover, the conflict involves people or communities to whom one has generally recognized obligations of kinship, friendship, membership, or alliance, it may be ethically improper—and, from some perspectives, politically culpable—for one to remain neutral. Thus, many sociologists of science were dismayed by their colleague Steve Fuller's recondite account of the comparable scientific status of evolutionary theory and Intelligent Design during his testimony in a school-board trial in the United States.[32] Similarly, many feminist scholars were disturbed by anthropologist Saba Mahmood's representation of Muslim women's self-subjection to patriarchal teachings as a mode of personal agency comparable to forms of agency valued by Western feminists (Mahmood 2005). There is, it might be said, a time for programmatic symmetry to be laid aside and for strong partisanship and explicit advocacy or critique to be taken up. I believe there are such times. But I also believe that they are not determinable in the abstract or in advance. Rather, I would say, those times are determined for each of us by the relevant particulars of our personal histories, identities, and obligations as well as by the particulars of the conditions that present themselves. Here, as often elsewhere, the best—most ethically responsive and intellectually responsible—way to handle the apparent difficulties created by relativist commitments is to relativize even further—that is, to acknowledge the significance, for oneself as for others, of even broader ranges and more subtle forms of contingent circumstances.

It may be presumed that, in their public actions as scholars, both Fuller and Mahmood were responsive to the relevance of various contingent circumstances. There is room for argument, however, about their respective decisions and also reason to speculate about how they arrived at them. It is not clear, for example, that Fuller considered as carefully as he might have done the long-range intellectual, educational, and political consequences of his testimony, as a credentialed social scientist, on behalf of the crypto-creationist side in the school-board trial. Similarly, it is not clear that Mahmood presented all the crucially relevant features of the lives and situations of the Muslim women she studied—for example, the history or threat of physical violence under which they may have acted. One may also wonder to what extent Fuller's generally populist sentiments, often directed against the science establishment, put him in ideological alignment with the promoters of Intelligent Design.[33] And similarly, one may wonder to what extent Mahmood's strict neutrality in her representation of the Egyptian women's mosque movement reflected a degree of protectiveness toward Islam, which, as a Pakistani-born woman (and in view of ongoing displays of Western arrogance and condescension), she could share with her subjects. In short, it is not clear that these two seeming demonstrations of objectionable relativistic impartiality were actually altogether impartial. Indeed, what may disturb critics of relativism in such cases is not politically objectionable neutrality but evidence of a relevant bias, muffled by a claim or show of fairness, for what is viewed as the wrong side of some current political struggle.

The preceding discussion acknowledges the genuine difficulties that may be presented in ethically or politically charged situations by a cultivated consciousness of the dependence of our perceptions, interpretations, and judgments on culturally and historically variable perspectives—or what are often called relativistic views. But I want to say something against the facile and, I think, fundamentally improper association of such a consciousness with ethically irresponsible or politically culpable quietism. Clearly, the ways we act politically, the forms taken by our partisanship and by either our advocacy or our critiques, can be determined more rather than less thoughtfully and on the basis of information that is more rather than less extensive, accurate, and relevant. Political actions can also be determined with greater rather than lesser concern for possible consequences, for broader and longer-term rather than only immediate and local consequences, and for consequences for wider rather than narrower ranges of people. In all these respects, political actions can be judged better rather than worse in the sense of being more rather than less ethically responsive and more rather than less effective in achieving either the particular political ends sought or some more broadly shared social goals. For these reasons, relativistic views, in the senses I have evoked here, do not make ethical or political judgments impossible. Moreover, when political activities are assessed in terms of the broad ethical and pragmatic dimensions just described, their energies are not diminished by what are called relativist convictions. On the contrary, it seems obvious that such convictions—that is, an acute consciousness of the historical and cultural contingency of human perceptions, interpretations, and judgments (including one's own) and of the sometimes significant variability of human interests and perspectives—would tend to make someone's political activities both more ethically responsive than and at least as effective as actions undertaken by someone with resolutely universalist, absolutist, and/or objectivist convictions regarding the Truth and the Way.

Relativistic considerations do not commonly paralyze personal agency. They may, however, affect the form of the actions one takes and the processes by which one arrives at them. For example, a strong consciousness of the possible relevance of unknown conditions and alternative perspectives may qualify the terms and tones in which one issues a denunciation or calls for an intervention. Such considerations may also make one hesitate—take more time, review a wider range of options—before one grabs a gun or gives an order to fire one. It is not clear, however, that these are politically undesirable effects. As the sorry history of many political movements and interventions suggests, the contingencies we deny and the variability we overlook for the sake of solidarity or for a show of unshakable conviction commonly come back to hit us or haunt us.

When thoughtfully worked through and put into practice responsibly, relativist convictions—in the sense of a cultivated consciousness of the variability and contingency of what operates as the real, the true, or the good—are neither ethically nor politically compromising. But, of course, not all relativistic convictions *are* thoughtfully worked through or responsibly evoked. On the contrary, expressed in sloganized forms (*everybody has his own opinion, who's to say what's good or bad*, and so forth), such ideas can be voiced mindlessly, lazily, and often with very bad manners (for example, condescendingly) or with very base motives (for example, to justify otherwise objectionable self-serving policies or practices). This brings me to some concluding observations.

IV

There are many reasons why the invocation of relativism, or even just reference to it as a topic, can be distasteful. For one thing, its exceptional elusiveness, as detailed earlier, makes relativism a genuine headache to think about, difficult to describe coherently, and almost impossible to argue about productively. Also, the recurrent philosophical equation of relativism with foolish or crude positions, such as an everything-is-equal-to-everything-else egalitarianism or an anything-goes nihilism, operates as a distinct disincentive to introduction of the term, even where it would be descriptively apt. Most significantly, perhaps, the mindless, lazy, or cynical voicing of relativist slogans, as just described, gives relativism an understandably bad name among intellectually and ethically scrupulous people and, for that reason and others, leads to its strenuous disavowal by thinkers whose views, given a range of familiar characterizations, might well be regarded as relativistic.[34] If we seek to dispel fire-breathing chimeras, these difficulties must be recognized.

General observations to the effect that meanings and values are radically contingent, or that perceptions, interpretations, and judgments are essentially variable, have been articulated with a variety of affects and motives: earnestly, ironically, in good faith, in bad faith, despairingly, gleefully, to challenge dubious claims of objectivity or universality, to explain incomprehensible difference, to plead for tolerance, to justify neglect. But occasional or even frequent delivery with bad manners or base motives does not make such observations invalid. Nor does it oblige us to shore up or reaffirm otherwise dubious conceptions of objective truth, universal value, or transcendent criteria. Rather, when general ideas of variability or contingency are invoked and applied objectionably—for instance, where an ill-considered view is lazily excused as "just one man's opinion" or an immediately consequential objectionable practice is shrugged off as "traditional in our/their culture"—then exposure and criticism are properly directed at the laziness, cynicism, or obscurantism involved, not at general observations of variability or contingency, or at that cloudy creature "relativism."

Finally, the horror or embarrassment of relativism is, I suspect, the horror or humiliation of mortality. Observations of radical contingency and irreducible variability are disagreeable because they remind us that our achievements are fragile, that our meanings are not altogether under our control, and that there may not be truth at the end of our efforts or justice at the end of our struggles. They are disagreeable because they oblige us to recognize the limited significance of all that we hold important, the perishability of all that we cherish, and our own fickleness and faithlessness. Some people regard these reminders as pessimistic or nihilistic; others see them as useful in pursuit of a sensible and ethical life. It's no doubt a matter of personal temperament. As folk-relativist wisdom has it, it takes all kinds to make a world and there's no point arguing about tastes. Of course that would include intellectual worlds and philosophical tastes.*

---

* "The Chimera of Relativism: A Tragicomedy" originally appeared in *Common Knowledge*, vol. 17:1, pp. 13-26. © 2011, Duke University Press. All rights reserved. Republished by permission of the copyright holder. www.dukeupress.edu

Chapter Three

Religion, Science, and the Humanities: An Interview

This exchange was originally posted on the Social Science Research Council blog, *The Immanent Frame*, by Nathan Schneider, an editor of the blog. The heading read as follows: "Barbara Herrnstein Smith is a distinguished literary scholar at both Brown and Duke who, since her undergraduate days, has had a special interest in the uses and misuses of scientific psychology. Her latest book, which stems from her 2006 Terry Lectures at Yale University, is *Natural Reflections: Human Cognition at the Nexus of Science and Religion* (Yale, 2009). It explores the ways in which contemporary cognitive science and evolutionary psychology are being called upon to, once and for all, explain religion." (Schneider 2010)

NS: *Natural Reflections* has been the subject of a lively debate on Stanley Fish's blog at *The New York Times*. Have you found the exchange productive?

BHS: One-shot retorts, or seesaw exchanges on blogs, are rarely models of intellectually productive discussion, but Stanley Fish's columns attract thoughtful readers, and I found the responses to his column on *Natural Reflections* instructive. Two related anxieties were repeatedly voiced on the basis of Fish's description of my evenhanded—or, in fact, determinedly symmetrical—treatment of religious beliefs and what we take as scientific knowledge. One is that I am flattening out important differences between them. The other is that I'm refusing to take a stand on a major issue of our time, and thus—wittingly or unwittingly—giving aid and comfort to the wrong side. The first of these worries is unwarranted. While I locate the differences between "science" and "religion" on multiple levels, I don't diminish either the significance of such differences or the stakes that may be involved in identifying them accurately. The second worry is, I think, misplaced in principle, and reflects increasingly oversimplified public views of science, religion, and the relations between them. Most of the commentators anxious about what side the book comes out on are concerned, I think, about such issues as the promotion of creationist ideas in science classes, or the clerical condemnation of contraceptive devices or homosexuality—that is, public issues in which noisy literalist convictions clash with established scientific accounts, or where informed secular attitudes are confronted by uncompromising ecclesiastic doctrine. Such concerns are understandable and I share them. But taking a clear stand on such issues does not require choosing sides between Science and Religion, conceived as monolithic adversaries in an epic battle.

NS: How does this kind of discussion compare with what your previous books have generated?

BHS: Though I think of myself as a peace-loving scholar, what you call "lively debate" seems to be my destiny—or, perhaps, addiction. Virtually all my books have been involved in lively enough intellectual clashes: value wars, theory wars, culture wars, and science wars, among others. In the late 1980s, at the height of the so-called canon wars, a writer for *The New York Times* described *Contingencies of Value* as "a bible of relativism"—and he didn't mean it as a compliment. Indeed, it was my initially surprised encounter with such overheated reactions to my account of literary value that led me to think more closely about such head-on intellectual collisions—what I came to call "the microdynamics of incommensurability." I describe how they play out in current debates over belief, knowledge, truth, and science in two subsequent books, *Belief and Resistance* and *Scandalous Knowledge*.

NS: Well before the publication of the book, videos of the Terry Lectures on which it is based were available online. Did responses that you received from the public or other scholars on the basis of those videos affect how the book developed?

BHS: Some people who watched the videos told me about it and murmured general appreciations, but what affected the development of *Natural Reflections* most significantly were the responses of students to presentations of my views in seminars that I gave at Duke and Brown while turning the lectures into a book. The groups included, at various times, the daughter of a rabbi, two strenuous secularists from abroad, a devotee of Daniel Dennett, and at least five people reexamining their relation to the Catholic Church. I found, without quite planning it, that my efforts in each session were directed in large measure to keeping everyone on board—engaged, active, talking, and thinking, rather than grandstanding or sulking—as we went through the readings. By the conclusion of each semester, there seemed to be a way of putting things—of describing and understanding the nature of "religion," "science," "belief," and the relations among them—that was acceptable to virtually everyone (though I lost the rabbi's daughter early on, and one of the hardline secularists held out to the bitter end). It was the process of reaching that way of putting things, and especially discovering what made it go well or badly, that was crucial for what I came to see and want as the ethos of the book.

NS: What first brought you, as a scholar of literature, to matters of science and religion?

BHS: As it happens, one of my first published works (in a magazine of undergraduate writing at City College in New York in the 1950s) was on the psychology of religious conversion. I was much taken by William James's *Varieties of Religious Experience*, though I also kept a copy of Nietzsche's *Zarathustra* in my pocket. Matters of science, particularly biology and psychology, occupied me centrally during my school years and, though I went on to receive degrees in literature and have worked in such fields as Renaissance poetry and critical theory, those interests have remained strong and are reflected in virtually everything I have written. As for matters of religion, they were there all along, though I didn't always identify them as such. Of course, when Renaissance poetry isn't about love, it's about religion. My experiences studying and teaching works such as Donne's *Holy Sonnets* and, for several years, Milton's *Paradise Lost* stood me in good stead in my recent encounters with Christian theology, contemporary biblical exegesis, and the complexities of religious sentiment. Also, critical theory has always been concerned with the nature of truth and the operations of rhetoric, imagination, illusion, and belief—think of Aristotle's *Poetics*—all questions that are central as well to the study of religion.

NS: *The Times* posts mention your once having worked with the great psychologist B. F. Skinner. Did he influence you, and has his influence borne itself out in your career?

BHS: The answer is yes, but it needs some context. What brought me into Skinner's orbit was not behaviorism. It was a summer job as a technician in his laboratory when I was already a student in literature at Brandeis. But I learned a lot about behaviorism along the way—certainly enough to know that it was not manifestly absurd or Satanic. Most significantly, as it turned out, I had the chance to read Skinner's *Verbal Behavior* in manuscript—which is to say, before Noam Chomsky's review of it. Few people have actually read the book, which has nothing to do with rats, pigeons, or Pavlovian-conditioned children. In any case, the account Skinner develops there helped form my sense of language-use as a dynamic, embodied, context-sensitive social practice. Other major influences on that view were books I was reading around that time by anthropological linguist Benjamin Lee Whorf and literary theorist Kenneth Burke—both homegrown American originals, I might note, like Skinner himself. All of this inclined me to be skeptical of what I saw as Chomsky's impoverished conception of language and implausible account of how it is acquired and used—and also, for related reasons, of Habermas's notion of communication ethics. These and other skepticisms deriving from, among other things, my undergraduate work on the psychology of perception, early encounters with William James and Dewey, and that pocketful of Nietzsche put me at odds for the next fifty years with widely held views in language theory, value theory, epistemology, and, most relevantly for this conversation, what is now called cognitive science.

NS: What, in particular, about recent cognitive science of religion caught your attention?

BHS: In spite of the skepticisms just mentioned, I approached works such as Lawson and McCauley's *Rethinking Religion*, Pascal Boyer's *Religion Explained*, and Scott Atran's *In Gods We Trust* with considerable interest, viewing them initially as contemporary continuations and updates of the great naturalistic tradition in the study of religion—works by figures such as Hume, Weber, and Durkheim. The up-front association with evolutionary theory was intriguing, and I hoped to find out what was new in both anthropology and religious studies, and perhaps have something to report about it all for my Terry Lectures. In the course of working through several shelves of volumes and numerous articles and reviews, I learned quite a bit about the institutional politics of religious studies and about some exceedingly bemusing beliefs and practices exhibited by people around the globe. But what ended up engaging my attention most significantly were the no less bemusing beliefs and practices exhibited by these contemporary researchers of religion themselves. So, I decided to frame my report on these developments from my perspective as a part-time sociologist of knowledge and to include, in my assessment of the now self-dubbed "cognitive science of religion" (I call it the New Naturalism), some duly critical and cautionary observations.

NS: How much do you think cognitive science, when shed of its more ideological exaggerations, can really tell us about religion?

BHS: What seems right to me is the idea that many widespread and recurrent types of belief and practice associated with religion reflect the operation of quite general human cognitive and behavioral tendencies. That idea doesn't originate, of course, with cognitive science. Expressed in different terms, we find it in Hume's *Natural History of Religion* and the work of many later theorists of religion. What's more original in the new approach is the idea that many of those tendencies reflect the evolutionary history of the species. To the extent that the cognitive science of religion elaborates those ideas and connects them to other ongoing work on religion, culture, cognition, and human behavior, its contributions can be substantial. What seems dubious to me is the claim that the tendencies in question reflect the activation of specific Stone Age mental mechanisms that can be, or already have been, identified by cognitive scientists. What seems utterly stultifying is the attendant suggestion that everything else said about religion is irrelevant, superficial, or pre-scientific.

NS: How much does this New Naturalism share with the New Atheism of Richard Dawkins and, for instance, Daniel Dennett, whose *Breaking the Spell* calls for a new, naturalistic science of religion?

BHS: It's important not to confuse the project I refer to as the New Naturalism—that is, cognitive-evolutionary studies of religion—with the so-called New Atheism. The New Naturalists are attempting to explain religion; the New Atheists are seeking to discredit it. Not all New Naturalists are atheists, and the project does not arise from an antipathy to religion. Boyer, a cheerful Frenchman, generally maintains an anthropologist's neutral distance from the beliefs he describes. Atran, a serious American, expresses an ambivalent appreciation of religion throughout his book. Dennett's efforts in *Breaking the Spell*, which would qualify as New Naturalist as well as New Atheist, are limited in both regards by a very narrow understanding of religion and a correspondingly dubious conception of beliefs—religious and otherwise—as static, discrete items of cerebral furniture.

NS: By holding up the work of classicist Walter Burkert above cognitive scientists like Scott Atran and Pascal Boyer, are you arguing that scientists should leave explanations of religion to humanists?

BHS: Not at all. Burkert's *Creation of the Sacred: Tracks of Biology in Early Religions* figures in my book, not as a humanistic explanation of religion, but as an evolutionary-biological explanation of it that is duly historically informed and otherwise intellectually spacious—as many New Naturalist explanations are not. My point is not that humanists can explain religion (or anything else) better than scientists but that, if your objective is to develop empirically responsive, intellectually connectible naturalistic accounts of religion, then the resources of humanistic scholarship—including its archives, objects of study, participant perspectives, and techniques of analysis—should be recognized as valuable and necessary ingredients. Burkert's explanations of various features of religion—he deals with sacrifice, oracles, priests, prayer, moral commandments, and many other things—are often more compelling than Boyer's or Atran's not because they're softer or sweeter but because, among other things, they are better grounded empirically. All three invoke evolutionary biology, primate studies, genetics, and game theory. Typically, however, Boyer's and Atran's explanations come down to speculations, offered as facts and findings, about hypothetical, unobservable mental mechanisms. Burkert's accounts come down to observations (and, to be sure, also speculations) about recurrent patterns of human behavior as evidenced in manuscripts, inscriptions, historical records, and archeological artifacts. Of course Burkert's experience as a scholar of ancient civilizations probably made him especially attentive to political and institutional aspects of religion, and also to imaginative elaborations of religious beliefs and practices, all of which are significantly neglected in New Naturalist accounts. But such experience need not be confined to humanists. It is available to anthropologists and psychologists if they think it is significant for the project at hand. After all, Burkert, undertaking pretty much the same project as Boyer and Atran, made himself familiar with a considerable array of new research and theory in the sciences before offering an account of the psycho-biological springs of religion. The trouble is not the cognitive scientists' limited knowledge of art, literature, or political and social history, but their failure to grant the relevance of such fields of knowledge to the ongoing project of "explaining religion."

NS: How compelling do you find the recent trend among those trying to bring science to bear on literary theory? Should humanists generally be striving to draw more from the "hard" sciences in their work?

BHS: The question is apt. I've been thinking about that "trend" (as you put it) quite a bit lately, and I hope to write about it. I'm always mindful of my own early participation in such efforts—for example, in *Poetic Closure*, by making use of gestalt psychology to describe the perception and experience of literary forms. My models at the time—admirable ones, I still think—were E. H. Gombrich's *Art and Illusion* and Leonard B. Meyer's *Emotion and Meaning in Music*. I'm certainly sympathetic to projects involving interdisciplinary incorporations and extensions and could point to an array of achievements, current as well as past, that attest to their value. Work by a number of Duke colleagues comes to mind (Mark Hansen and Robert Mitchell, among others), along with Elizabeth Wilson's recent book, *Psychosomatic*. As such work illustrates, relevantly informed scholars in literary studies and other humanities-based disciplines may incorporate concepts and findings from natural-science fields in ways that can be subtle, original, genuinely illuminating, and sometimes significantly transformative for their own fields. Burkert's *Creation of the Sacred* is, of course, another example. I would have to add, however, that some of the current efforts to bring science (under some very limited views of it) into literary studies are energized by extremely dubious aims and motives. I think especially of hapless offerings by people who are persuaded that their discipline has gone to the dogs (the major alleged agent of that dissolution being some vague menace called "postmodernism") and who think it can be redeemed only by large, duly stiffening, injections of natural science. These convictions have a surprising degree of uptake among minimally informed people outside the field of literary studies, including, I'm sorry to see, some distinguished scientists.

NS: How can humanists—particularly after incidents like the Sokal Affair—make their voices heard by those in the scientific community in a productive way?

BHS: Alan Sokal is not a good representative of the scientific community in that regard, but his hoax was an effective piece of mischief that did much to deepen an already existing chasm created by a century of mutual ignorance and mutual caricature. I would stress the mutuality of that ignorance and those caricatures. Humanists and scientists are inevitably divided by significant—and, I think, by no means undesirable—differences of intellectual training, intellectual temperament, and intellectual idiom. They can converse productively with each other, however, when both recognize their own limits and provincialisms, and when each grants due respect to the worthiness of the other's projects and achievements. The readiness of some publicly visible scientists to dismiss humanities scholarship as trivial or unenlightened is, of course, painful. But if humanists, including scholars of religion, seek to be heard across the two-culture divide, they must be willing to give ear to reports of relevant developments in the natural sciences and to acknowledge—and, I would say, challenge—the readiness of many of their colleagues to cast scientists in correspondingly demoting, not to say demonizing, roles.

Chapter Four

Anthropotheology: Bruno Latour Speaking Religiously

> *To talk about religion again. ... No one appointed him, nothing marked him out, if not the certainty that once we modify, as he has done (as he thinks he has done), the common version of the sciences, everything else can start to change—first and foremost, religion.*
>
> —Latour, *Rejoicing: Or The Torments of Religious Speech*

Bruno Latour is a relatively recent taste in the Anglo-American academy. He has been publishing important work in the anthropology and/or sociology of science since the 1980s but, until the past five years or so, has been greeted largely with antagonism or indifference by science and humanities faculty alike. While Latour's work (or tendentiously selected passages from it) was a prime target of science warriors in the 1990s, people in the humanities have generally found his writings too remote from current concerns to seem interesting (he has had little to say, for example, about the politics of race or gender, at least explicitly) or too closely associated with the natural sciences to seem approachable. In recent years, however, invocations of ideas and approaches associated with Latour have become commonplace, along with citations of specific texts he has authored, especially those with irresistible titles.

For many readers, Latour is most closely identified with actor-network theory (ANT), a set of radical concepts and sophisticated methods developed originally in the sociology of science. He is also well known, especially among people in the humanities, as a subtle analyst of modernity and, more generally, as a vigorous advocate of environmentalism. Less widely known are Latour's extensive writings on religion. These include, from the 1970s, a doctoral dissertation on biblical interpretation and a related study of the early twentieth-century writer, Charles Péguy; a long essay from 1996, *Petite réflexion sur le culte modern des dieux Faitiches,* later translated as "On the Cult of the Factish Gods"; an important lecture from 2002, "'Thou Shall Not Freeze-Frame,' or, How Not to Misunderstand the Science and Religion Debate;" and, also from 2002, a small but in many ways extraordinary book, *Jubiler ou Les tourments de la parole religieuse,* recently translated as *Rejoicing: Or the Torments of Religious Speech.* Religion, or religious "being" as a specific mode of existence, figures centrally in *An Inquiry into Modes of Existence* and, along with Nature, is one of the major categories of analysis in Latour's Gifford lectures, *Facing Gaia: Eight Lectures on the New Climatic Regime* (2017). Indeed, the hope and effort to frame a proper and—in Latour's important term—"diplomatic" account of religion, and especially of its relation to science, have been central motivating forces in his work for at least the past two decades and, in some respects, from the beginning.

Virtually any reader who undertakes the serious study of Latour's writings (as distinct from casual sampling or heresy-hunting) will find them engrossing, instructive, often exhilarating and always impressive. But "the humanities" make up a very mixed package of practices in the present Anglo-American academy, and people currently working in the fields so designated make up a very mixed multitude. The ways in which any of us take up Latour's work, to "recompose" that package or otherwise, will depend, of course, on our particular assessments of those practices and on our aims and angles more generally.[35] Additionally, because attempting to do things "with Latour" will, sooner or later, involve encounters with his religious writings and with their particular concerns and perspectives, the ways we take up his work are also likely to depend on what he would call our "attachments." A detailed examination of Latour's writings on religion is beyond the scope of this essay. What I hope to do here is suggest the interest of these writings for scholars in the humanities and also to indicate the ways in which they seem likely to create problems for such readers, including or perhaps especially for longtime admirers of his work.

I

When readers fail to understand why I have continually changed fields, and when they do not see the overall logic of my research ... their comments amuse me, for I know of no other author who has so stubbornly pursued the same research project for 25 years, day after day, while filling up the same files in response to the same sets of questions.

—Latour, "Biography of an Inquiry"

Alluding to his successive studies of science, art, politics, and law, Latour has described his general project as "the comparative study of the various ways in which the central institutions of our cultures produce truth" or, as he also calls those ways, "truth regimes" (2009, ix.) The regime on which his earlier work focuses is that of the modern natural sciences. In empirical—archival and onsite—investigations conducted in the late 1970s and early '80s and in their theoretical elaborations as actor-network theory, Latour has sought to demonstrate that what are commonly taken as scientific truths—facts, laws, discoveries, entities—are not, as commonly assumed, fixed, prior, and given by "nature" (itself radically reconceptualized by Latour) but, rather, the contingent products of dynamic networks of multiple, heterogeneous elements. The elements include both humans—scientists, technicians, bureaucrats, and sometimes farmers or fishermen—and nonhuman agents or actors, from sick cows and virulent microbes to pulleys and petri dishes. All these are moving in different, potentially conflicting directions, and some are stronger or weaker than others; but, in laboratories and other centers of calculation and control, some elements can be linked together to form associations that are effective in serving particular human ends. It is the pragmatically effective linking of such elements that secures what we call the truth of scientific facts (for example, the microbe theory of disease or the structure of DNA) and that sustains what we experience as the reality of the entities associated with those facts (for example, microbes or genes).[36]

This constructivist-pragmatist understanding of scientific truth and knowledge reflects an increasingly commanding tradition of research and theory that extends from Ludwik Fleck's *Genesis and Development of a Scientific Fact,* originally published in 1935, to the writings of a number of mid-twentieth-century historians, sociologists, and philosophers of science and, from there, to ongoing work in the field now known as science and technology studies (STS). As formulated, elaborated, and promoted in writings by, among others, Latour, it has proved compelling to increasing numbers of humanities scholars, along with researchers and theorists in the social sciences, both as a set of conceptual and methodological resources for work in their own fields and also as a well-developed alternative to still-dominant positivist views. As research and teaching in the humanities continue to involve closer connections to the natural sciences, Latour's work in this tradition can be especially important and, in regard to the earnest or aggressive scientism sometimes displayed in these developments,[37] it can be especially instructive. There has been no radical break between Latour's early and more recent work on science and no reversal in the direction of his thought. Since his "coming out as a philosopher" (Latour 2010), however, he has supplemented and, in some crucial regards, sought to supersede ANT and empirical science studies more generally with an array of speculative methods and explicitly metaphysical projects. He has also been increasingly explicit about what he evokes, especially in *Rejoicing*, as his particular task or responsibility: that is, to read aright the texts and inscriptions of the religion that, as he says, "matters" to him and to translate, transmit, and make effective its message for those he calls "Moderns."[38] Latour's thirty-year-long comparative investigation of truth-regimes was pursued in good measure in the service of that task. *An Inquiry into Modes of Existence* can be seen as the consummation of the investigation and, with the Gifford lectures, as his most valiant venture to date as missionary to the Moderns.

II

Is existence not among the perfections indispensable for respect, which the idea of belief never allows us to preserve? Thus I had to come back to the crack that runs between epistemological questions and ontological questions. The new history of the sciences has allowed me to slip in between the two.
   —Latour, "On the Cult of the Factish Gods"

Contrary to routine misunderstandings of constructivist accounts of scientific facts, to be constructed—made, built, fabricated, put together from heterogeneous elements—is not to be unreal. Abstract facts, like material artifacts, are assembled and composed, but both are "real" in the sense of being, at least provisionally, stable and consequential. The same can be said of gods and other religious beings: demons and divinities, spirits and fetishes.

As Latour tells the story in "On the Cult of the Factish Gods" (2010), Gold Coast natives, scorned by European traders and invaders, insisted that certain wooden dolls—dubbed *fétiches* by the Portuguese—were gods. The natives, Latour observes, had "constructed" something "that went beyond them." But, he asks, is this not true as well of the facts constructed by Western scientists, for example, Louis Pasteur's "ferment of lactic acid," the existence of which emerges through laboratory instruments and tests? (16) Moderns, with all the apparatus of scientific rationality, no less than supposedly primitive people with their wooden divinities, invest things that they themselves have made with a power that goes beyond them. Facts and fetishes, demons and ferments: "All ask to exist," Latour writes. "None is caught in the choice ... between construction and reality, but each requires particular forms of existence whose list of specifications must be carefully drawn up" (45).

Fetish-gods, like scientific facts, acquire their potency—or, as it may be called, their "truth" or their "reality"—within a framework of specific ideas, habits, discourses, and material apparatus; but the potency of neither can survive outside those frameworks. Whether divinities or DNA molecules (and, as Latour extends the point in *Modes of Existence*, whether cats, mats, machines, political collectives, or characters in novels), the conditions of their continued existence—he calls them "felicity conditions"—are highly specific, not always in place, and always more or less fragile. In the case of religious icons, for example, they are breakable by the acts of impassioned iconoclasts or modern "critical thinkers."[39]

The imputation of an equivalent real existence to the facts of modern science and the divinities of putatively primitive religions—or, put differently, the acknowledgment of their equivalent ontological status—is an example of what Latour calls "symmetrical anthropology." He explains its method and aim in the essay: "By taking the most respected beings of a culture—our own—as examples, we can shed light on the most despised beings of another culture" (45). The most respected beings of our own culture are scientifically established facts and entities. The most despised beings are African fetish-gods, the demons afflicting the immigrant patients of a French ethnopsychiatrist and, not quite "of another culture," the Virgin as sighted at Lourdes. Latour's symmetrical anthropology can be seen as due scientific impartiality, as a generous exercise of the sympathetic imagination, or, perhaps, as practicing relativism with a vengeance. It can also be seen as a sophisticated elaboration of the rhetorical move known, especially in theological circles and in response to derisive iconoclasms, as *tu quoque:* "You, too! the supposedly enlightened ones: you do just what you scorn us, the supposedly benighted ones, for doing."

In describing the facts of modern science symmetrically with religious beings, Latour does not seek to demote the authority of the truth-regime of Western science. What he seeks to demote—indeed, to undo utterly—is a set of dichotomies and commonly skewed dualisms that have become central to modern Western thought: nature as divided from society, objects as divided from subjects, real as opposed to manmade or constructed, and existent as opposed to (merely) believed-in. But of course, and not incidentally, he thereby promotes the epistemic dignity of the experiences of those who fear demons or see visions of the Virgin, and the ontological dignity of those beings themselves.

In a classic constructivist treatment, our experience of the truth of scientific facts and the reality of visions of divinities would be understood in terms of more general, largely social-psychological dynamics. Thus Fleck, in *Genesis and Development of a Scientific Fact,* describes the complex processes involved in the formation and stabilization of what he calls "belief systems," with scientific paradigms and religious doctrines, along with political and other ideologies, as examples. In Fleck's account, the coherence and stability of all such systems are preserved through the ongoing mutual adjustment of the perceptions, prior beliefs, background assumptions, and shared material practices of the interacting members of a social group or "thought collective." Fleck called the resulting shared sense of the truth of some fact or doctrine among the members of such a group a "harmony of illusions"—illusions not in the familiar and itself dubious sense that there was some otherwise verifiable set of objective facts that contradicted them, but insofar as that sense of truth was projected outward and regarded as an objective correspondence of idea and world.

Latour has repeatedly expressed admiration for Fleck's work, and the affinities of their respective accounts of facts and truth are evident.[40] The detailed historical-sociological narrative of the establishment of the microbe theory of disease in *The Pasteurization of France* closely parallels Fleck's narrative, in *Genesis and Development,* of the establishment of the Wassermann test for syphilis, including the way a key pathogen is coaxed into existence in the laboratory. Crucial to Fleck's account, however, is an analysis of the social-psychological dynamics involved whereas Latour rejects explanatory appeals to the psychological and, in *Modes of Existence,* banishes the term "belief."[41] Also, significantly, while both reject table-thumping empiricisms in favor of constructivist understandings of facts and truths, Latour invokes a rather obscurely defined "second empiricism" to ground the ontologies of *Modes of Existence*. These differences—as much matters of intellectual project and genre as of philosophical position—mark an important space between the tradition of science studies with which Latour's work has been associated and his more recent writings, those on religion and more generally.

III

There exists a form of original utterance that speaks of the present, of definitive presence, of completion, of the fulfillment of time … ; a form of speech whose sole characteristic is to constitute those it is addressed to as being close and saved; a kind of vehicle that differs absolutely from those we've evolved elsewhere to accede to the distant in order to control information about the world.
—Latour, *Rejoicing*

The perennially disputed relation between the truths of science and those of religion is addressed directly in Latour's essay, "'Thou Shall Not Freeze-Frame,' or, How Not to Misunderstand the Science and Religion Debate" (2005). Originally a talk for a lecture series titled "Science, Religion and the Human Experience," the essay offers a set of formulations regarding that relation that Latour develops in detail in *Rejoicing* and iterates in more recent writings. The essay also involves, contra iconoclasts of all persuasions, a crucially revised interpretation of the biblical commandment prohibiting images. Rhetorically reflexive throughout, the essay is, among other things, a mock (but not mocked) sermon. Latour writes: "Religion, at least in the tradition I am going to talk from, namely the Christian one, is a way of preaching, of predicating, of enunciating truth in a certain manner—this is why I have to mimic in writing the situation of an oration given from the pulpit" (28).

Latour begins with a strong contrast between "speaking religiously," evidently as in prayer or ritual utterance, and what he calls "double-click communication," that is, the idea or ideal of an unmediated transfer of information. The truth of a double-click message, if any such existed, would be its exact correspondence to an objectively determinable state of affairs. Religious speech acts, on the other hand, "transport" not information but persons. In religious speech as in love talk, what attests to the truth of an utterance is not its correspondence to some putatively objective reality but its renewal of speakers' and hearers' confidence in the reality of something vital: a sense of closeness; a promise of futurity (29-31). Here as elsewhere in Latour's writings on religion, claims are put forth largely through analogy, allusion, and intimation—which is not untypical, of course, of theological arguments or sermons.

Clearly, Latour observes, it would be improper, what he calls a "category mistake," to judge the truth of a religious speech act using double-click communication as a measure.[42] Just as it would be wrong to maintain that sentences such as "I love you" have no truth value just because they possess no informational content, it is wrong, in seeking to understand the angel Gabriel's salutation to the Virgin, to ask who Mary was, to ponder "whether or not she was really a Virgin," or to imagine that she might have been impregnated with "spermatic rays." "Paradoxically," Latour writes, "by formatting questions in the procrustean bed of information transfer so as to get at 'exactly' what it meant, I would have *deformed* it, transmogrified it into an absurd belief, the sort of belief that weighs religion down and lets it slide toward the refuse heap of past obscurantism" (33). In *Rejoicing*, Latour describes—at length and with considerable scorn—religious scholars' efforts to explicate New Testament texts so as to make them more reasonable-sounding, more conformant to historical data, or otherwise palatable to intellectual tastes corrupted, as he sees it, by Double Click (here and elsewhere personified and often associated ironically, or maybe not so ironically, with "the Evil One"). He continues in the essay: "The only way to understand stories such as that of the Annunciation is to *repeat* them, that is to utter again a Word which produces into the listener the same *effect*," one that "impregnates... with the same gift, the same present of renewed presence. Tonight, I am your Gabriel!" (33).

Seeking explicitly to evoke the power and effects of religious transmission, Latour turns from verbal to visual representation and comments on a set of strong images from Christian iconography. We do not, or should not, assess such images, he observes, by their fidelity to presumed true originals. Nor should we isolate or "freeze-frame" them from the flow of mediating representations that enable their truths to be realized (this being Latour's revision of the second commandment). He goes on to stress the comparably vital role of relays of inscriptions, images, and other representations in science (reports, charts, photographs, mathematical formulae, and so forth). "Truth," Latour writes, "is not to be found in correspondence—either between the word and the world in the case of science, or between the original and the copy in the case of religion—but in taking up again the task of *continuing* the flow, of elongating the cascade of mediations one step further" (46). The commonly supposed objective realities behind genes or the microbe theory of disease are like the mistakenly supposed "originals" of visual representations of the empty Sepulcher or of the arresting thorn-crowned face of Jesus in a trompe-l'oeil painting of the Veronica veil. In all these, what matters, what sustains the truth of the events and the reality of the figures in question, is the continuity of the practices of representation that mediate their existence.

Elaborating these points in the essay's concluding pages, Latour observes, in what operates as an important and continuing distinction, that, while the mediating chains of reference that secure the truths of science are counterparts to the flows of utterances and images that convey the truths of religion, the relays in each go "in two different directions" (46). In science, they bring what is far close (for example, through astronomical photographs, charts, and models), but religious texts and images bring us to what is near—our neighbor and our salvation. Because Moderns have worshipped the false idol of Double Click, Latour maintains, they have misunderstood—indeed, reversed—how truth and reality are secured both in science and in religion. To correct what he calls this "comedy of errors," he offers a set of alternative characterizations of religious belief and scientific knowledge that are central to *Modes of Existence* and repeated, with variations, in his Gifford lectures. "Belief," he writes, "is not a quasi-knowledge question *plus* a leap of faith to reach even *further* away; knowledge is not a quasi-belief question that would be answerable by looking directly at things close at hand." Rather, a leap of religious faith "aims at jumping, dancing towards the present and the close, to redirect attention away from indifference and habituation." Conversely but comparably, knowledge in science "is not a direct grasp of the plain and the visible ... but an extraordinarily daring, complex, and intricate confidence in chains of nested transformations of documents that, through many different types of proofs, lead toward new types of visions that force us to break away from the intuitions and prejudices of common sense" (45-46). The parallels and reversals in this set of comparisons are striking. Simultaneously vague and enthusiastic, they join an evocation of the most familiar and accessible experiences of religious faith to a celebration of the most heroic activities and exalted achievements of science while maintaining a sharp distinction between the two. They are nothing if not diplomatic.

IV

In seeking to frame an account of the relations between science and religion that is both generally acceptable and also corrective of what he sees as past philosophical and theological errors, Latour has taken on a task that is immense and, as suggested in *Rejoicing*, variously—certainly rhetorically and perhaps, for Latour, conceptually as well—"tormented." Such an account must negotiate steep differences of view between Moderns, many of them invested in conventionally celebratory views of science and some of them scornfully antireligious, and Christian communicants, many of them invested in conventionally orthodox religious views and some of them resentfully anti-science. Thus, while secular-minded readers may welcome a theology that claims neither supernatural nor substantial status for its god(s) and that segregates religion from politics and morality, communicants might feel that something essential has been lost in the negotiations. Accordingly, Latour's accounts of religion vis-à-vis science operate with a good bit of euphemism, circumlocution, studied vagueness, and, it could be said, equivocation. For example, while Latour derides familiar theological allusions to realms "above" or "beyond" the natural or the material, the apparent heterodox force of such gestures is considerably defused by his equally strong efforts to undermine familiar understandings of "nature" and "matter." Similarly, while he seems to suggest that religion is immanence all the way down and all the way up, too, it is not surprising that fellow faithful sense in his texts assurances of something like orthodoxy.[43]

To speak religiously to Moderns, Latour has tied together a theoretically sophisticated account of scientific knowledge with a rhetorically deft Christian apologetics to forge a singular quasi-symmetrical anthropotheology. The writings that compose it are bold, inventive, and in many ways compelling. Structurally and stylistically, *Rejoicing*, *Modes of Existence*, and related essays are remarkable works of lyric philosophizing, recalling works by Kierkegaard and, in their strong personal voice, Nietzsche. Fellow theologians are likely to be most appreciative of the originality of their formulations and also most closely attuned to their distinctive idioms.[44] Other readers will find them a rich resource for ongoing, reprised, or newly conceived scholarly projects. Historians and theorists of Western modernity will profitably engage with Latour's theologically inflected takes on law, politics, and economics. Those in literary and visual studies will appreciate his suggestive accounts of the re-presencing effects of texts and images, religious and otherwise. And humanities scholars of all stripes will be delighted by passages of an order of wit and literateness—vernacular as well as erudite—not often encountered in the pages of theologians, not to mention social scientists. Readers and scholars in all these fields, however, are likely to be perplexed by various aspects of these writings and to find them, to various extents, intellectually or experientially alien.

V

> *I am not going to speak of religion in general, as if there existed some universal domain, topic, or problem called "religion" that could allow one to compare divinities, rituals, and beliefs from Papua New Guinea to Mecca, from Easter Island to Vatican City. A person of faith has only one religion, as a child has only one mother.*
>
> —Latour, "'Thou Shall Not Freeze-Frame'"

In his writings on religion, Latour has been concerned with a relatively confined set of aspects of a vast and multifaceted subject.[45] The focus is on religious representation and utterance, which, in *Rejoicing* and related essays, are identified largely with Christian iconography, New Testament texts, and the verbal practices of Catholic communicants. In *Modes of Existence*, the religious mode of existence is explicitly restricted to the beings of Christianity while demons, ghosts, fetish-gods, and other exotic divinities are assigned to a separate, somewhat obscurely described mode labeled "metamorphosis." Beings of the latter kind are sustained not, as in religion-proper,[46] by flows of sacred texts and images, but by a process that Latour calls "psychogenesis"—associated with shamans, exorcism, psychotropic drugs, and psychoanalysis—and explains as "the exterior production of interiorities." Also, strikingly, no other major religious tradition is mentioned in *Modes of Existence*, his five-hundred-page-plus "Anthropology of the Moderns." Writing as a professed Catholic, Latour could not be expected to deal with other faiths in the same manner or detail as he deals with Christianity. Nevertheless, readers are likely to miss some acknowledgment of the existence of other religious traditions and also of their variety, both as observed and as experienced.[47]

Experience carries a great deal of weight in *Modes of Existence*. The inquiry's method, "a second empiricism," is, Latour explains, a developed or extreme version of William James's "radical empiricism": that is, the inclusion of nothing that is *not* in experience and the exclusion of nothing that *is*.[48] Moreover, the test of the truth of its accounts of Modern values is, he tells readers, the accord of those accounts with their own experience. With regard to religion, however, the appeals to experience are highly selective and readers may find them otherwise thorny.

Some of the difficulties can be seen in the following passages, in which Latour specifies the mode of existence of "religious beings"—that is, the beings of Christianity, also identified as "the beings sensitive to the Word"—and explains their categorical differences from what he calls "the beings of metamorphosis."

> Religious beings ... are truly beings; there's really no reason to doubt this. They come from outside, they grip us, dwell in us, talk to us, invite us; we address them, pray to them, beseech them.
>
> By granting them their own ontological status, we can already advance quite far in our respect for experience. We shall no longer have to deny thousands of years of testimony; we shall no longer need to assert sanctimoniously that all the prophets, all the martyrs, all the exegetes, all the faithful have "deceived themselves" in "mistaking" for real beings what were "in fact nothing but" words or brain waves. ...
>
> It appears infinitely simpler, more economical, more elegant, too, to stick to the testimony of the saints, the mystics, the confessors, and the faithful, in order to direct our attention toward *that toward which* they direct theirs: beings come to them and demand that they be instituted by them. But these beings have the peculiar feature of *appearing* to those whose souls they overwhelm in saving them. ... If we are to be empirical, then, these are the ones we must follow. ... Like the beings of metamorphosis, religious beings belong to a genre "susceptible to being turned on and off." With one difference: if they appear—and our cities and countrysides are still dotted with sanctuaries erected to harbor the emotions these apparitions have aroused—they *disappear* even more surely. Moreover, this intermittence has provided the basis for mockery, and has been taken as proof of their lack of being ...; the critical spirit has not held back in this regard. But the big advantage of an inquiry into modes of existence is that it can, on the contrary, *include* this feature in the specifications: one

of the characteristics of religious beings is that *neither their appearance nor their disappearance can be controlled.* (308-309)

This seems to be saying that the existence of the beings of religion-proper is (only) in the particular experiences of those who experience such beings and that the reality of their existence is secured by our agreeing—out of respect for those experiences—not to question that reality. It also seems to be saying that, in spite of evident similarities, the invisible beings proper to Christianity cannot exist in the same manner as the invisible beings of other religions because only the former conform to what Christianity teaches about such beings.[49] The advantage noted here ("the big advantage of an inquiry into modes of existence is that it can ... *include* this feature in the specifications") is that the person conducting such an inquiry can specify as a singular feature of the ontology of the beings of his own religion—and, indeed, as a manifestation of their autonomous power (that is, to appear and disappear *uncontrollably*)—what might otherwise be taken as their compromised reality: that is, the nondemonstrability of their existence and the fitfulness of their presence even in the experience of the faithful.

There is, clearly, no arguing with the structure or elements of an ontological claim of this kind. Readers not party to the type of stipulative logic involved may feel there is something hocus-pocus about it or note the apparent self-affirming circularity. Latour, however, defends its rationality strenuously: "I hope the reader will do me justice on this point: not once in this inquiry have I required anyone to give up the most ordinary logic; I have only asked that, with the *same* ordinary reasoning, the same natural language, they follow *other* threads. ... [The beings of religion] are rational through and through. Like psyches. Like fictions. Like references" (307). And, in any case, one may find it hard not to be charmed by a universe emptied of "matter" and animated by invisible "beings" flitting among souls, sliding among the pages of old books, in company with Heathcliff and perhaps Athena, as real as quarks and as reasonable as cats or mats.

VI

If I still dare speak, it's only because I think I can brush aside the shadow that the ways of science once cast over the ways of being produced by religion.

—Latour, *Rejoicing*

Labeled an "inquiry," Modes of Existence can be seen as Latour's final report on his thirty-year-long comparative investigation of truth-regimes. Aspects of the inquiry, however, clearly had foregone conclusions, a number of which operate as axioms or, in Latour's term, "pre[-]positions," that is, as proper attitudes taken or given in advance. The sharp distinction and mutual incommensurability of the modes of veridiction of science and religion appear to be axioms of this kind. As set forth in Modes of Existence, these features obtain across the board: all truth-regimes involve distinct modes of existence, which themselves involve distinct discursive tonalities, interpretive keys, and modes of veridiction. One of the work's central conclusions (or givens), however, is that Moderns have brought much unhappiness upon themselves, the rest of humanity, and the rest of creation through their confusion of the truth-regimes associated with science and religion *in particular* and through their failure to respect the differences between the respective interpretive keys and tonalities of each.

One may agree: there is something tone-deaf in seeking to establish the truth of the Annunciation the way one might that of a theory of biological evolution. One can also see the broad advantages of maintaining a clear distinction between the modes of veridiction associated with religions and the natural sciences: it protects visions of the Virgin from dismissal in terms of empirical facticity and evidence regarding Jupiter's moons from dismissal in terms of scriptural or ecclesiastical authority. Indeed, a strict partition of "science" and "religion" has obvious benefits for both, as demonstrated by the recurrent efforts of advocates or defenders of each to establish one.[50] Nevertheless, in view of the close, extensive, and formative connections between the development of the modern Western sciences and the institutions of religion, one must question the extent to which their respective discursive tonalities or even truth-regimes can be distinguished, certainly historically and, in some regards, currently as well.[51] And, in view of the exceptionally heterogeneous and continuously shifting contents of the packages of ideas, practices, institutions, and communities that have been and could be assembled under each of these terms, "science" as well as "religion" (and the latter even if confined to Christianity), one must question the conceptual coherence and practical workability of any claim about the fundamental nature of either of them or of their relationship.

When Moderns "start talking about the 'conflict between Science and Religion,'" Latour writes, "they act as though it were a matter of opposing (or 'reconciling,' which is worse) two types of approach: one that would give us Matter, the 'here below,' the rational, the natural, and one that would offer us the spiritual, the beyond, the supernatural, the supreme values!" (322). The sort of opposition and/or reconciliation Latour describes here is familiar in the idea of "nonoverlapping magisteria," as proposed by biologist Stephen Jay Gould (1999). In Gould's division, authority over the realm of facts and accounts of the natural world is claimed for science while religion is granted authority over the realm of values along with instruction in moral conduct. Gould's apportionment of the epistemic and moral universe is endorsed by many scientists, who believe they have the best of the bargain, and is also accepted by many theologians, happy to be granted clear title to a piece of the territory.[52] Of course, partitions like Gould's perpetuate what Latour identifies as key problematic dualisms of Modern thought: facts and values, matter and spirit, nature and culture. But Latour has sought only to challenge the terms in which those partitions have been drawn, not their existence as such. Few contemporary theorists have been more alert to the problems of conceptual segregation than Latour or devoted as much energy to exposing the dubious divides of Western thought. To the extent, however, that *Modes of Existence* depicts "science" and "religion" as distinct and counterpoised monoliths, its revisionist ontology, even as it discards familiar dualisms or significantly redistributes their traditionally defining elements, goes some distance toward perpetuating one of the most dubious of them.

VII

I've got better things to do than to portray the ups and downs of the children of last century: things like altering the arrow of progress[,] ... giving another meaning to the long history of the West, doing away with modernization.

—Latour, *Rejoicing*

Latour does not claim to be a historian, but his work involves a good bit of historiography as well as important theorizing about historicity and temporality.[53] *The Pasteurization of France* is, among other things, a history of the emergence of modern theories of disease; *We Have Never Been Modern* is, of course, a thorough overturning of modernity's self-flattering autobiography; *Rejoicing* relates the successive efforts of Christian theologians to meet the successive challenges of rationalism, both classical and modern; and both *Modes of Existence* and *Facing Gaia* involve significantly revised versions of major chapters of Western social, political, and intellectual history.

The fields and approaches that make up science studies, including actor-network theory (ANT), are programmatically anti-whiggish. They reject familiar heroic-progressivist narratives of the history of science and comparable manifest-destiny accounts of the history of technology. Moreover, they tell very different kinds of stories about both. ANT's defining method is the slow, careful tracing of the construction of contingent networks of multiple, heterogeneous, complexly interrelated elements. While ANT accounts register practical successes and failures, they do not score the ideas and artifacts whose construction they narrate as intrinsically grand or foolish, nor do they portray the human agents whose efforts they follow as blind or faithful to (the) truth.

In his role of missionary to the Moderns, Latour sets aside this commitment to symmetrical historiography. Seeking to "[alter] the arrow of progress," he flips it around to point backward. Where Latour's Moderns tell of a rise from darkness and superstition through Reason and Science, he tells of a fall from unity and faith through the embrace of those very (misunderstood) values. His tale is of a community assembled by a salvific message; of the entrance of malign forces offering knowledge and power; of the folly and fumbles of leaders; of a message obscured, a people left wandering, and a land in ruin; and of the chance, perhaps, of redemption and renewal.[54]

The tale is old and familiar. To be sure, that is no reason to dismiss it. Nevertheless, the idea of the Scientific Revolution and the European Enlightenment as catastrophes for humanity is likely to be resisted by many of Latour's academic readers, including—and in spite of their shared sense of the ills of modernity—a good number of them in the humanities. It is not that such readers endorse familiar celebratory accounts. Intellectual, literary, and social historians, along with political theorists, are more likely to regard both developments, along with the Protestant Reformation, the Industrial Revolution, and other chapters in standard histories of modernity, as very mixed bags with very complex and variously operating ingredients. It is, rather, that they have learned to be skeptical of myths of a Fall, whether into Technology, Commerce, Individualism, or Fragmentation, and also of moralized histories, whether triumphal or nostalgic.[55] Many of us are inclined to see not only the twentieth century but also the past two millennia and perhaps the entire history of humanity as a long series of, precisely, "ups and downs": of local gains and losses, dominances and defeats; of emergences and extinctions both large and small; but not of globally grand triumphs and/or great botches, in either order. And many of us find the idea of modernity, or "the secular age," or any age, as a "parenthesis" in human history (Latour 2004, 234)—as if an interruption or aberration—very peculiar. "But, of course!" Latour might exclaim. "That is because you are Moderns—or worse, Postmoderns!"

It is true: many of us are, to various extents, one or the other of these or both of them. Insofar as we are, even as we appreciate and appropriate Latour's work around the clock, we will be troubled and more or less alienated by an image of the West in which critical thought is cast as the enemy of the ways to truth and the fools and knaves of intellectual history are named Galileo, Hume, Kant, Voltaire, Émile Durkheim, Sigmund Freud, and Jacques Derrida.

VIII

Psychology is to the subject what epistemology is to the object. One must be countered as forcefully as the other in order for experience to be tracked.

—Latour, *Modes of Existence*

In spite of the homage that he pays to William James, Latour rejects the relevance of psychology to the understanding of experience, including religious experience.[56] The field or, rather, fields of psychology (there are, of course, many specializations and variants) have a lot to answer for in the way of simplistic accounts of, among other things, the nature, sources, and effects of religious beliefs and experiences. There are, however, quarters of those fields where the assumptions of classic epistemology are rejected as strenuously as Latour rejects them (and for many of the same reasons) and where questions of subjects, psyches, and persons are approached in ways that accord closely with his own elaborated views of them. The relevant approaches, called, variously, "nonrepresentational," "ecological," "embodied," or "enactive," also suggest ways to understand beliefs—religious, scientific, and other—that do justice to their complex phenomenological dynamics.[57]

Latour is comparably insistent that the religious mode of existence, and religion as such (or at least Christianity), cannot be approached by the social sciences more generally:

> There is a risk, obviously, that [the] requirement to treat religion rationally will be mistaken for a return to the critical spirit, that is, to the good old "good sense" of the social sciences. But it should be clear by now that we can expect nothing at all from the "social explanation" of religion, which would amount to losing the thread of the salvation-bearers by breaking it and replacing it with another, while seeking to prove that "behind" religion there is, for example, "society," "carefully concealed" but "reversed" and "disguised." Such an "explanation" would amount to losing religion. ... There is nothing "behind" religion ... since each mode is its own explanation, complete in its kind. (*Modes of Existence*, 307)

But, of course, religion, religious experience, and the related operations of mediation that Latour describes in ontological terms can be and have been described otherwise, by no means always either reductively or critically. Ethnologically and historically informed accounts of religious ideas, practices, and institutions, Christian and other, along with subtle explorations of religious subjectivities, have been produced for more than a century by anthropologists, classicists, and other scholars of religion who have shown no interest in exposing anything "behind" the objects of their study or inclination to mock anything within them.

If Latour makes little use of these accounts, it is not because he is unaware of them. It is because they are irrelevant to what he has taken to be his task. For Latour, to "speak well" of religion—that is, of Christianity—is to speak of it religiously, which means in its traditional scriptural, theological, and homiletic idioms *and not otherwise*. The propriety or cordiality required here is less a matter of language than of attitude and, indeed, of attachment. The attitudes and attachments of a person of faith occupying the role of communicant or theological apologist are crucially different from those of a scholar of religion comparing practices from Papua New Guinea to Vatican City (though they may, in fact, be the same person). *Emic* and *etic,* inside and outside, the experienced and the observed: the differences between them cannot be bridged; they can only be finessed. This, the "hard problem" of the philosophical tradition, is also the hard problem of anthropotheological diplomacy.[58] As simultaneously anthropologist of and missionary to the Moderns, Latour has attempted to solve or negotiate it by forging an original idiom—a way of speaking—that joins compelling evocations of religious experience to passionate theorizing in the service of a prophetic summons to worldwide conversion. There is good reason to think the mission will fail. What has been constructed along the way, however, will reward our exploration for some time to come.

Chapter Five

What Was "Close Reading"?: A Century of Method in Literary Studies

This chapter originated as a talk to a group of young digital humanities scholars seeking to learn about the history of methods in the humanities. I am not a historian by profession and it is not my usual role, but I attempt here to mobilize my sixty years in and around the literary academy to address a double question: how did "close reading" figure in Anglo-American literary studies over the course of the past century, and how does it figure now in the discourses of the digital humanities? At the end, I offer some general reflections on methods, past and possibly future, in literary studies.

I

Close reading, it has been said, is the "primary methodology" of literary studies (Jockers 2013, 6). That is true in a sense, but the term methodology suggests something more coherent, circumscribed, and specifically research-focused than has been the case here. Reading individual texts with attention to their linguistic features and rhetorical operations is very different, of course, from subjecting large bodies of digitized materials to the sorts of computational processes now wittily (and mischievously) called "distant reading." But the term *close reading* has been used to name some very diverse activities: from a New Critic unraveling Shakespearian puns in the 1930s to a Marxist scholar exposing the political unconscious of Victorian novels in the 1980s or, today, a first-grader "analyzing" a book by Dr. Seuss in accord with the directives of the American national Common Core Curriculum.

If not quite a methodology, however, the practices of close reading have certainly been a persistent feature of Anglo-American literary studies. Indeed, their ongoing performance may be the one constant in a field notorious for its succession of new "approaches." The parade is familiar from textbook rubrics: the old historical philology followed by the New Criticism, structuralism, reader-response criticism, New Historicism, feminist criticism, deconstruction, cultural studies, ideology critique, and so forth, with many others in between. Almost any parade becomes comical when it goes on long enough, and the succession of approaches in literary studies has been seen as showing the futility of all of them. There is good reason to reject that view, as I shall suggest below. But the point I want to stress here is that, for all the shifts they reflect, every one of the approaches just named—from the New Criticism, and indeed from the old historical philology, through deconstruction, to ideology critique—involved reading individual texts closely. The texts varied from presumed literary masterpieces to works of popular culture and documents of manifest oppression; the discourses that directed their examination varied from Christian humanism to structuralist linguistics to queer theory; and the spirit in which they were examined varied from the appreciative to the disinterested to the deeply suspicious. Nevertheless, throughout the twentieth century, whatever the mood, motive, or materials, if one was teaching literature or doing literary criticism in the Anglo-American literary academy, one was likely to be reading at least some individual texts closely.

If, as seems to be the case, the practices of close reading have operated in literary studies not as one method among others but as virtually definitive of the field, then how are we to understand a method whose advocates define it in opposition to—and, indeed, as superseding—precisely those practices? Depending on one's perspective, the ascendance of "distant reading" can be seen as marking the dissolution of literary studies, at least as a humanities discipline (see, for example, Kirsch 2014), or as the proper elevation of the field into computational posthumanism. In arguments defending either of these two opposed views, invocations of "close reading"—celebratory or derogatory—are a recurrent ploy. I consider some of these invocations below. First, however, a bit of history will be instructive.

II

The term *close reading* refers not only to an activity with regard to texts but also to a type of text itself: a technically informed, fine-grained analysis of some piece of writing, usually in connection with some broader question of interest. The practice has multiple ancestors, including classical rhetorical analysis, biblical exegesis, and legal interpretation, and it also has some cousins, such as iconology and psychoanalysis. All of these would have been familiar to the small group of accomplished British dons and poets whose efforts to reform literary study in the 1920s and 1930s came to be called "the New Criticism" and whose critical essays served as models for the practices that came to be called "close reading."

In its time, the New Criticism was young, cool, and radical. I. A. Richards was a beginning instructor at Cambridge University when he assigned the classroom exercises in poetry reading that led to his influential book *Practical Criticism*, published in 1929. William Empson, a student of Richards, was twenty-three when he wrote *Seven Types of Ambiguity* (1947), which became a paradigm of virtuoso close reading for several generations. T. S. Eliot's *Sacred Wood: Essays on Poetry and Criticism*, published in 1920, excoriated dull scholarship from a position of confident connoisseurship and set the standard for high-toned literary discriminations.

The views and practices of this group were introduced to American readers in the 1930s by the southern poet and critic John Crowe Ransom, and they were taken up with particular enthusiasm by two of his students, Cleanth Brooks and Robert Penn Warren, originally at Vanderbilt, later at Yale. Brooks and Warren's coauthored textbook, *Understanding Poetry*, first published in 1938, was immensely influential. By the time a second edition appeared in 1950, it was being assigned in more than 250 colleges and universities in the United States (Davis 2011). Its chapter titles—"Structure," "Tension," "Irony," "Imagery," "Rhythm," "Tone"— are the familiar obsessions of New Critical close readings, and its large collection of examples and "texts for further study" weighed heavily in the academic canon for decades afterward: Shakespeare's sonnets, Keats's odes, works by Browning, Tennyson, and Gerard Manley Hopkins, and modernist works by, among others, Ezra Pound, Elizabeth Bishop, and William Carlos Williams.

The New Criticism was promoted as a corrective for the pedantries of early twentieth-century literary scholarship. But the practices of close reading themselves were promoted as a specifically pedagogical remedy for what were seen as the inadequacies of college students: young men (almost exclusively—we are speaking of the 1920s and 1930s) who, in Brooks' words, "actually approached Keats's *Ode to a Nightingale* in the same spirit and with the same expectations with which they approached an editorial in the local county newspaper or an advertisement in the current Sears, Roebuck catalogue" (1979, 593).[59] Indeed, the popularity and persistence of those practices have often been attributed, not altogether unjustly, to their handiness for teachers. Some forty years after the publication of *Understanding Poetry,* Hugh Kenner, a learned critic of modernist poetry, remarked: "The curious thing is how a classroom strategy could come to mistake itself for a critical discipline" (1976, 36; cited by Brooks 1995, 84). Kenner was being snide, of course, but the association of close reading with pedagogy—as distinct from learned criticism—persisted through much of the century.

To return to the history: Before the New Criticism, in the first quarter of the century, the study of literature consisted largely of the production, transmission, and acquisition of facts about sets of texts. What one established as a scholar, imparted as a teacher, and learned as a student were commonly the names of historically important authors and some basic facts about their lives; the titles, publication dates, and sources—especially classical—of their major works; relations of influence among them; and the readily observable features that distinguished forms, styles, and genres (the medieval romance, the Petrarchan sonnet, the Jacobean drama, and so forth). One could say that, before the New Criticism, literary study was "distant reading" with a vengeance. With the work of the New Critics, it moved increasingly from filling library shelves with scholarly editions and literary histories to studying and describing how individual texts produced the effects that gave them historical importance or current interest.

Against the historians and philologists, who treated literary texts as dusty achievements, the New Critics stressed literature as art.[60] A poem, wrote Ransom, is a "living object" (1937, 601). As poets and wordsmiths themselves, they were interested in the craft of text making. As men of letters (Cambridge, Vanderbilt, Yale) and, in increasing numbers since the 1960s, literary women (the earliest included poet Josephine Miles at Berkeley), they were appreciators of aesthetic effects. Though close reading is often described as a method of interpretation, the New Critics—certainly the first generation of them—were concerned less with establishing the meaning of a text than with understanding its operative machinery. Indeed, a New Critical "reading" was something like an exercise in reverse engineering: the examination of an artifact to see how it was made and how it worked.

The aims and spirit of the New Criticism are well represented in an inaugural piece by Ransom, published in 1937, titled "Criticism, Inc." He explains the title: "Professors of literature are learned but not critical men. … Nevertheless it is from the professors of literature … that I should hope eventually for the erection of intelligent standards of criticism. It is their business. … Perhaps I use a distasteful figure, but I have the idea that what we need is Criticism, Inc., or Criticism, Ltd" (587–88). The "figure" or metaphor—literary studies as a business—would have been "distasteful" because literary academics understood that art and letters were remote from commerce. That shared understanding, and the shared antagonism to Big Business and Big Industry, created a bond and otherwise strange fellowship between the conservative southern poets and the left-wing New York intellectuals who, in the 1940s and 1950s, were the major advocates and practitioners of the New Criticism in the United States.

Ransom, referring in the essay to the relation between historical scholarship and criticism, is clear about which should be up front and on top: "Behind appreciation … is historical scholarship. It is indispensable. But it is instrumental and cannot be the end itself. In this respect historical studies have the same standing as linguistic studies [that is, philology]: language and history are aids" (595). He also makes explicit the New Critics' pedagogical mission: "The students of the future must be permitted to study literature, and not merely about literature. But I think this is what the good students have always wanted to do. The wonder is that they have allowed themselves so long to be denied" (588).

Only toward the end of the essay does Ransom say what the new forms of literary study would consist in, and then in just a few phrases:

"They would be technical studies of poetry[;] for instance, [in a particular poem] ... if they treated its metric; its inversions ...; its tropes; its fictions, or inventions by which it secures 'aesthetic distance' ...; or any other devices, on the general understanding that any systematic usage which does not hold good for prose is a poetic device" (600). This is followed by a bit of New Critical mystery mongering:

> The critic should regard the poem as nothing short of a desperate ontological or metaphysical manoeuvre. ... [The] poem celebrates the object which is real, individual, and qualitatively infinite. [The critic] knows that his practical interests will reduce this living object to a mere utility, and that his sciences will disintegrate it for their convenience into their respective abstracts. The poet wishes to defend his object's existence against its enemies, and the critic wishes to know what he is doing, and how. (601)

As I said: rather like reverse engineering.

It was often recalled that Ransom, like Richards before him, had summoned criticism to be more "scientific," but his meaning and aim were often misunderstood. Here is the passage:

> Criticism must become more scientific, or precise and systematic, and this means that it must be developed by the collective and sustained effort of learned persons—which means that its proper seat is in the universities. ... It will never be a very exact science, or even a nearly exact one. But neither will psychology, if that term continues to refer to psychic rather than physical phenomena; nor will sociology ...; nor even will economics. It does not matter whether we call them sciences or just systematic studies. ...The[y] ... have immeasurably improved in understanding since they were taken over by the universities, and the same career looks possible for criticism. (Ransom 1937, 587–88)

As I think is clear, Ransom was not seeking to make literary criticism scientific in the positivist sense of the era. What he sought, rather, was to move it from the margins of the field to the center and to claim for its practitioners a form of technical expertise that elevated its status from an occasional pursuit to a duly accredited and properly housed academic program. If that sounds like what promoters of the digital humanities are seeking today, both for themselves and for the practices of "distant reading," then it is very much to the point here, with all the historical ironies that doubling involves.

Other doublings and ironies are apparent in the hostile reactions that greeted the New Criticism. Such reactions came, as Ransom anticipated, from "the present incumbents of the professorial chairs," who, he wrote derisively, "spend a lifetime in compiling the data of literature and yet rarely or never commit to a literary judgment" (587). A notable counterpolemic appeared in 1949 in the pages of *PMLA*. Its author was Douglas Bush, an eminent Shakespeare scholar at Harvard. Its self-consciously conservative title was "The New Criticism: Some Old-Fashioned Queries." Bush writes: "The new criticism, the offspring of Mr. Richards and Mr. Eliot, has carried the marks of a mixed heredity. ... [Their] close reading of poetry has braced the flaccid sinews of this generation of readers and has had some highly beneficial effects upon teaching and writing ... . But ... a scholar-historian may not be disposed to grant all its claims and assumptions" (13). The grudging concessions and sense of strained dignity are familiar in the responses of upstaged eminences. By the late 1940s the hold of the old historicism had been effectively shaken, the literary academy was an increasingly lively place, and Bush could speak ruefully but accurately of the New Critics' "tone of conscious superiority" (14). After chiding Ransom and others for gaffes reflecting their supposed ignorance of history, he goes on to charge them with leaving ordinary readers out in the cold: "The common reader might go so far as to think that poetry deals with life .... However valuable the processes and results of the new criticism, for some readers its preoccupation with technique, its aloof intellectuality, its fear of emotion and action, its avoidance of moral values, ... all this suggests the dangers of a timid aestheticism" (20). Bush concludes his essay with a bit of table-turning: the charge—or, rather, countercharge—of scientism: "Since poetry does after all deal with experience, the most fastidious critics have to touch on it [;] yet they may give the impression that they are looking, not at human beings, but at specimens mounted on slides. Indeed, though the critics have censured scholarship for aping science, their own aims and methods seem much more deserving of the charge" (20).

Much of this is quaint but also, I think, remarkably recognizable. Virtually the same set of general claims and charges were rehearsed throughout the century, irrespective of the methods at issue and, indeed, irrespective of which side was being taken, revolutionary or counterrevolutionary. Thirty or forty years later, the same resentful tones would be heard from the New Critics themselves, no longer young, or cool, or radical and now upstaged by deconstruction and what they called "high theory." Ransom's reference to the vested interests of incumbent professors; the charge of elitism in Bush's invocation of "the common reader"; the exchanged charges of hyper-formalism (or "aestheticism") and scientism: these would all recur in the "theory wars" of the 1970s and 1980s, in the "canon wars" of the 1980s and 1990s, and, like much else in both these essays, in current polemics around the digital humanities. It looks like a pattern. It may even be a law of history. It is time to turn to Franco Moretti and the twenty-first century.

III

It was Moretti, of course, who broached the idea that literary studies would benefit from a turn away from close reading to a new set of practices that could be called "distant reading." The key passage appears in an essay titled "Conjectures on World Literature," originally published in 2000:

> The trouble with close reading (in all of its incarnations, from the new criticism to deconstruction) is that it necessarily depends on an extremely small canon .... And if you want to look beyond the canon ... close reading will not do it. It's not designed to do it, it's designed to do the opposite. At bottom, it's a theological exercise—very solemn treatment of very few texts taken very seriously—whereas what we really need is a little pact with the devil: we know how to read texts, now let's learn how *not* to read them. Distant reading: where distance, let me repeat it, *is a condition of knowledge*: it allows you to focus on units that are much smaller or much larger than the

text: devices, themes, tropes—or genres and systems. And if, between the very small and the very large, the text itself disappears, well, it is one of those cases when one can justifiably say, Less is more. If we want to understand the system in its entirety, we must accept losing something. We always pay a price for theoretical knowledge: reality is infinitely rich; concepts are abstract, are poor. But it's precisely this "poverty" that makes it possible to handle them, and therefore to know. This is why less is actually more. (Moretti 2000, 57–58)

The rhetoric is neat; the argument is bold; and several strong themes and claims echo in current arguments promoting the digital humanities. One is the theme of *scale:* the association of "close reading" with a few small things. Another is *knowledge:* the summons to literary critics and scholars to dare "to know," to produce more abstract, theoretical, and comprehensive knowledge. A third, the most consequential in institutional terms, is *less is more:* the suggestion that, for literary studies to operate as a field of genuine knowledge production, scholars must sacrifice something: their interest in "the text itself" and the centrality of close reading to their practices.

The themes of scale and less-is-more are sounded, less playfully, in an article by Matthew Wilkens titled "Canons, Close Reading, and the Evolution of Method" (2011). Wilkens sees literary canons as a product of the practices of close reading and also as something like a moral problem—a matter of injustice to the excluded—as well as a methodological one. He writes: "What little we do read is deeply nonrepresentative of the full field of literary and cultural production, as critics of our existing canons have rightly observed for decades …. So canons … are an enormous problem, one that follows from our single working method as literary scholars—that is, from the need to perform always and only close reading as a means of cultural analysis" (251). His solution is clear: "We need to do less close reading and more of anything and everything else that might help us extract information from and about texts as indicators of larger cultural issues" (251). After describing a computational study and the benefits of algorithmic methods, Wilkens concludes: "This [i.e., the turn to such methods] will hurt, but it will also result in categorically better, more broadly based, more inclusive, and finally more useful humanities scholarship" (257). The words are earnest, but the argument is dubious. Close reading is by no means the only working method of literary scholars. Moreover, while literary canons—that is, the works most widely celebrated and referenced in the culture, most regularly assigned in literature classes, and most frequently discussed in professional journals—commonly reflect, among other things, the tastes and interests of the professoriate, they are not determined by the reading practices of scholars. Most significantly, the mere existence of a literary canon places no limit on the number, the type, or the cultural status of the texts available for study, whether as cultural indicators or as anything else.

There are also the simple facts of the matter. As literary studies have been pursued under the auspices of structuralism, semiotics, New Historicism, deconstruction, feminism, critical race theory, postcolonial criticism, and queer theory, the types and cultural status of the texts examined by literary scholars *and* read closely in their classrooms have continuously expanded. Those "texts" now include writings of every form and provenance, whether currently admired or reviled and whether currently read or unread; and they can in principle (and often do in fact) include any inscription, or image, or artifact whatsoever. Wilkens suggests that less close reading of individual texts and more computational studies of large bodies of texts will somehow address the other "problem of the canon," by which he presumably means the past snubbing of popular writings and of works by women and members of minority groups. But if the offense is that many worthy or interesting texts remain unread because of past biases, then what is wanted, surely, is to have those texts *read,* not just counted. Wilkens seeks to promote research methods that he sees as undervalued in literary studies. The aim is commendable but his arguments do not serve it well.

The themes of *scale* and *knowledge* in Moretti's piece also recur in a book by Matthew Jockers titled *Macroanalysis: Digital Methods and Literary History* (2013). Jockers is explicit in associating desirable epistemic aims with the aims of science and proper research methods with the methods used by scientists. "The goal of science," he writes (he adds "we hope"), "is to develop the best possible explanation for some phenomenon. This is done via a careful and exhaustive gathering of evidence …. Literary studies should strive for a similar goal" (5–6). Like science, he continues, literary studies should seek the best methods available for gathering evidence and, again like science, should welcome big data and scale its methods accordingly.

It was Jockers' description of close reading as a "methodology" that I cited earlier. Here is the passage I was quoting: "The study of literature relies upon careful observation, the sustained, concentrated reading of text. This, our primary methodology, is 'close reading.' Science has a methodological advantage in the use of experimentation. Experimentation offers a method through which competing observations and conclusions may be tested and ruled out .... We are highly invested in interpretations, and it is very difficult to 'rule out' an interpretation" (6). Granting the value of literary interpretation in some regards, Jockers continues: "But interpretation is fueled by observation, and as a method of evidence gathering, observation ... is flawed. Despite all their efforts to repress them, researchers will have irrepressible biases .... Observation is flawed in the same way that generalization from the specific is flawed: ... the selection of the sample is always something less than perfect, and so the observed results are likewise imperfect" (6–7). Jockers believes that big data is solving these problems in the sciences. He writes: "Big data are fundamentally altering the way that much science and social science get done .... [M]any areas of research are no longer dependent upon controlled, artificial experiments or upon observations derived from data sampling" (7). And, he argues, big data should change the game in literary studies as well:

> Back in the 1990s, gathering literary evidence meant reading books, noting "things" (a phallic symbol here, a biblical reference there, a stylistic flourish, an allusion, and so on) and then interpreting: making sense and arguments out of those observations. Today, in the age of digital libraries and large-scale book-digitization projects, the nature of the "evidence" available to us has changed, radically .... [M]assive digital corpora offer us unprecedented access to the literary record and invite, even demand, a new type of evidence gathering and meaning making. The literary scholar of the twenty-first century can no longer be content with anecdotal evidence, with a few random "things" gathered from a few, even "representative,"

> texts. ... Like it or not, today's literary-historical scholar can no longer risk being just a close reader: the sheer quantity of available data makes the traditional practice of close reading untenable as an exhaustive or definitive method of evidence gathering. (8–9)

There is much to query here, about Jockers' view of science as well as of literary studies, but I will focus on a few central points. First, gathering evidence for claims is not a good way to describe research in most scientific fields, and generating interpretations by reading books and noting things randomly and foolishly is a singularly bad way to describe what literary scholars did in the 1990s or at any other time. More significantly here, close reading never figured in literary studies as a "definitive," much less "exhaustive," way to gather evidence. In the case of a critical study focused on the thought, style, or achievement of one or more particular authors, the close reading of texts written by those authors would have to be central to any claims made. But even there, and certainly where broader historical or cultural claims are involved, it would be remarkable for a literary scholar *not* to consult and invoke other documents and other types of data. The digital library is exceedingly handy for accessing such materials, but scholars' shelves have not been otherwise empty up to now. Jockers is saying that literary historians "can no longer risk being" something that they never have been: that is, "*just* close readers."

Second, the fact that some way of doing things is now possible does not make it necessary, as in Jockers' term "demand." Where big data is pertinent and computational processing would be useful, literary scholars should take advantage of both. But there is no research imperative built into the size of some potential data set. New methods enable new questions to be posed and old answers to be sharpened or corrected. But in any field of knowledge production, *significant* questions come out of ongoing interests and problems, not usually just methods as such. Digital libraries and powerful search engines have already become research tools for most literary scholars. Those who are still in a cave on these matters are well advised to become familiar with such resources and with the types of projects they enable. It is understandable that those already working on such projects would want to urge others to explore such resources as possible avenues to interesting new research for themselves.

But Jockers seems to be urging or cautioning more than that in his "like it or not" address to his colleagues. What people in literary studies are apt to hear him saying is that the existence of those "massive digital corpora" has made their current way of doing things inappropriate, impractical, and "untenable" and that they should be doing instead what he does.

A third point is important here. Jockers suggests that computational methods and big data allow researchers to avoid or mitigate the subjectivity and bias built into human observation and also what he calls the "flaw" of generalizing from less than exhaustive data sets. But the idea or ideal of objective observation, like that of complete data sets (in effect, inspecting every swan before concluding anything about the color of swans), has been effectively challenged by a century of empirical and theoretical work in the history, sociology, and philosophy of science. Jockers is not alone among contemporary literary scholars in his enthusiasm for, but rather old-fashioned view of, "science." (The persistent singular suggests a dubious conception of these heterogeneous practices as something monolithic.) For all the talk of "paradigm shifts" among digital humanists, literary Darwinists, and advocates of cognitive cultural studies, the notion of science to which they appeal tends to be fundamentally pre-Kuhnian.[61]

My allusion is, of course, to Thomas Kuhn, whose account of paradigm shifts in the history of chemistry and physics challenged simple progressivist views of how the natural sciences develop (Kuhn 1962). Even more pertinent here, however, is Paul Feyerabend's once scandalous, now celebrated, volume, published in 1975, titled *Against Method*. To the central question posed by philosophers of science—what method leads scientists to new discoveries and successful theories?—the prevailing answer was Karl Popper's "conjectures and refutations," or the so-called hypothetico-deductive method (Popper 1962).[62] Feyerabend's answer was, *Anything goes!* Contrary to outraged misreadings of his argument, he was saying not that scientists are capricious but that they are inventive, resourceful, and opportunistic (his examples included Galileo, Newton, and Einstein). Feyerabend's major point in *Against Method* is that there are no rules for success in science: that is, *no specifiable, generalizable way* to make discoveries or to produce persuasive, useable theories. The trouble with telling literary scholars to be "like scientists" in their methods is not that it is scientistic (though it is that as well) but that it tells them nothing in particular.

Jockers' worries over the methods used in literary studies are largely misplaced. The fact that a critic's interpretation is based in part on his or her observations as an individual reader does not compromise the interest or usefulness of that interpretation. The same is true of a scholar's historical account or theoretical claim. Of course that individuality—and inevitable, but not necessarily pernicious, bias—leaves the interpretation, claim, or account open to dispute by other readers or scholars. At the same time, however, the grounding in personal observation and experience opens the possibility of shareable insights and of connection to shareable experiences, which—largely, if not wholly—is what motivates our interest in a literary interpretation as such. And, along with connections to broader intellectual issues and other concerns, that grounding and that attendant possibility—of shareable insights and of connection to shareable experiences—are also what sustain the value of much historical and theoretical research in the humanities as such.

There is a long story to tell—longer than proper to claim space for here—about the differences of epistemic aim, achievement, and value in the humanities and the sciences.[63] But it is clear, I think, that a central question here is whether scholarly and critical methods in literary studies are properly assessed by reference to research methods in a field like geology—the specific science that Jockers invokes in an opening anecdote as having led, during a dinner conversation with a representative scientist, to his embarrassment at being unable to defend the ways in which research was done in his own field. More use of big data or computer algorithms by literary scholars will not solve the methodological problems that Jockers sees in literary studies. The problems, if that is what they are, are built into standard scholarly practice in the field, the legacy of classical humanistic practices reinforced by various contemporary approaches. The only way to end the embarrassment at what literary critics and literary historians do would be to persuade them to do something else instead, something more like social science research. That seems, in fact, to be what Moretti proposes in the name of "distant reading" and what Wilkens, Jockers, and others are promoting in the name of "digital humanities": not, perhaps, to cease doing traditional literary study altogether but to start doing less of it and to do more research in which the data consist of large numbers of texts and the questions asked and answered are independent of the interest—literary or otherwise—of any of those texts.

Less is more: perhaps. Moretti calls the exchange he proposes—less close reading for more genuine knowledge—"a little pact with the devil." Pacts with the devil, even what look like little ones, deserve scrutiny. Famously, one ends up getting less than one bargained for and losing more than one thought.

Moretti is right to describe close reading as "a theological exercise"; not, however, in the sense, as he suggests, of the reverential examination of a text but in the sense of an expansive commentary on one. In literary studies, a close reading, whether in a classroom or a journal article, is typically the occasion for more general observations and often for quite wide-ranging reflections. They may be observations about the style or genre of the text at hand, or about its author, or reflections on the era in which it was written. But they are often observations and reflections—more or less subtle, more or less original—about related human circumstances and experiences. And even when not especially subtle or original, they can afford some insight into, and a sense of connection with, the circumstances and experiences of people who are otherwise remote. Contrary to charges raised across the years, the practices promoted by the New Criticism did not require that readers "forget history" or "ignore the outside world." Full-dress close readings, now as ever, can be showy or strained. They can also be dim, thin, derivative, or pedestrian and, when motivated by a history of injury, sulky or venomous. But now as ever, they can offer those who hear or read them potentially illuminating engagements with regions of language, thought, and experience not otherwise commonly encountered.

"Close reading" makes a neat contrast to "distant reading," but not a pertinent one for promoting the value of computational methods for literary studies. My final example here is a pair of short passages in a coauthored book titled *Digital_Humanities*. Referring to the tools necessary "to thoughtfully and meaningfully sift through, analyze, visualize, map and evaluate the deluge of data … the digital age has unleashed," the authors write: "One way of navigating this process is through distant reading: a form of analysis that focuses on larger units. It is a term that is specifically arrayed against the deep hermeneutics of extracting meaning from a text through ever-closer, microscopic readings" (Burdick et al 2012, 39). A page later, they amplify this: "Close reading has its roots in the philological traditions of the humanities, but for more than a generation has often been equated with deep hermeneutics and exegesis, techniques in which interpretations are 'excavated' from a text through ever-closer readings of textual evidence, references, word choices, semantics, and registers" (40).

Both descriptions are awkwardly phrased, perhaps reflecting ideas inserted at various times by each of the five authors who collaborated in its production. But especially notable, I think, is the oddness of some of the language used to describe the activities associated with close reading. We generally speak of inferring or identifying meanings and of offering or suggesting interpretations, not of "extracting" or "excavating" them like teeth, oil, or corpses. Also odd are the phrases "ever-closer" and "ever-closer, microscopic." Close reading often involves attention to features such as word choice or, in connection with rhyme or alliteration, individual sounds or letters. But there is nothing like a microscope lens being focused "ever-closer," as if obsessively or maniacally. Like the rather violent "extract" or the creepy "excavate,"[64] such language makes the practices so described sound distinctly unpleasant, rather unnatural, and certainly very alien.

More significantly here and contrary to the contrast being drawn, attention to microfeatures does not distinguish the practices of close reading from those of distant reading or the digital humanities. We recall Moretti's inaugural description: "Distant reading ... allows you to focus on units that are *much smaller* or much larger than the text: *devices, themes, tropes*—or genres and systems" (emphasis added). And in fact, the tagging and counting of very small textual units figures centrally in many digital humanities projects, including studies by Moretti himself.

For example, in an article titled "Style, Inc. Reflections on Seven Thousand Titles (British Novels, 1740–1850)" (2009), Moretti displays the results of a study correlating the length, grammatical structure, and other features of those seven thousand titles with the number of novels published in each decade. He goes on to note historical trends in the occurrence of those features, including, in the later decades, when increasing numbers of novels flooded the market, the use of the indefinite article *(a)* versus the definite article *(the)*. This differential usage, he suggests, as in the titles *A Mummer's Wife* versus *The Infidel Father*, signaled to prospective consumers that a particular novel offered, respectively, either a progressive or a conservative point of view on social developments. Moretti elaborates the suggestion with considerable subtlety (noting, for example, that *a* is unrestricted and leaves the future open, whereas *the* is restricted and grammatically oriented to the past) and teases out the connotations of other microfeatures of those titles in an analysis that might have left Empson gasping with admiration.

Of course, Moretti's interpretation of the data he has gathered can be challenged. His account of the implications of *a* and *the* is based at least in part on his subjective observations, here as a practiced reader and writer of the English language, and different explanations of the historical trends he identifies could be and have been offered.[65] What is significant for the present argument is, first, that that would be true, mutatis mutandis, of the interpretation of the results of a study in any field, whatever the size of the data and however rigorous the computations: that is, the grounding of the interpretation in the inevitably limited observations of an individual human being. But, second, as I noted above, that kind of grounding is also what makes an interpretation potentially compelling and gives it interest for, and appropriability by, fellow human beings. You could see it as an exchange: less putative (and, I would say, spurious) objectivity for more actual interest and usefulness. A pretty good bargain, I think.

The law of literary history that Moretti discovers (or corroborates or illustrates) in the *Critical Inquiry* article is that stylistic features of literary texts—titles, in this case—reflect market forces. Hence the "Inc." in his own title, "Style, Inc.," which recalls the title of John Crowe Ransom's essay, "Criticism, Inc.," and may involve a glancing allusion. Ransom, we remember, was suggesting that criticism is the proper "business" of literary studies, and he noted that his readers might find the term "distasteful." Contrary to the old genteel-humanist view of art and letters as definitively remote from commerce, Moretti is suggesting that literature is a business like everything else. He means it provocatively, but his *Critical Inquiry* readers are not likely to find the association of art or letters with the market either distasteful or surprising. It has been some years—nearly a century—since literary studies was the "solemn treatment of very few texts."

IV

If you live a long time, you accumulate a lot of data. I offer here some general observations on method in literary studies. As I noted above, the succession of "approaches" in the field over the past century is not, in my view, a sign of futility or failure. What it reflects, rather, is a history of ongoing efforts by literary scholars to make their teaching and research responsive to developments within and beyond the academy, efforts that have been joined—and sometimes undercut—by the operation of dynamics that are general and recurrent enough to be called laws of history, at least of academic history. One such law is *Everybody always overdoes everything*. Professional pressures—inspiration energized by competition—push scholars into exhibiting the methods and virtues most conspicuously valued at the moment, and these tend be escalated and intensified to the point of exhaustion of the material, absurdity of the method, or pathology of the virtue.[66] A second law or set of forces, giving weight to perennial calls for reform or revolution, is the existence of chronic tensions between fundamental but opposed impulses and styles: for example, in literary studies, between populism and elitism, moralism—including political moralism—and formalism, and subjectivism and positivism. A third law follows from the interaction of the first two: *Everybody complains of being misrepresented and caricatured, and everybody is misrepresented and caricatured*. Bush complained of the New Critics' "cavalier" dismissal of historical scholarship;[67] and Ransom's treatment of it *was* pretty glib.

The developments to which scholars were responding during the twentieth century were quite significant. Literary study was one thing when a small number of Christian men were teaching the professionally aspiring sons of fellow professionals. It became another when members of an expanding professoriate were teaching students from middle- and working-class families or, later, when a sizeable number of faculty were women and a sizeable number of their students were from racial and ethnic minorities. And the field is yet another thing now, when faculty and students are more likely to encounter texts on screens than anywhere else, and everyone is scrambling for positions, funding, and status in a shrinking quarter of the academy.

Over the course of the century, in literary studies as in other fields, the responses to changing institutional conditions and to broader changes, including intellectual developments, commonly led not to revolutionary overturns but to shifts of relative dominance, and there were always extensive residues of prior practices. The field was never "taken over" by the New Criticism or by any other single movement, and it remained—not unhappily, I think—eclectic. The parade in the street was noisy and colorful, but up in the libraries, historians continued to write literary history, editors continued to produce editions of literary works, and, in both libraries and classrooms, almost everybody read at least some texts more or less closely.

Some changes of practice in literary studies proposed by advocates of the digital humanities *would* be revolutionary, and the developments they observe and cite as requiring these changes—from the technological to the neurological—are, in their eyes, no less so.[68] Talk of "seismic" and "tectonic" shifts is pervasive. Clearly, though, the interest and utility of close reading do not vanish in the face of digital libraries or ubiquitous computation. On the contrary, in the century upon us, where channels of communication are not only increasingly computerized but also increasingly corporatized and where texts of all kinds are turned to manipulative ends with digitally multiplied effectiveness, the ability and disposition to read texts attentively, one by one (in addition, of course, to digital sophistication), is likely to be an advantage. That textual ability and disposition also remains more generally crucial. The hope of receiving such reading is what keeps most of us—scholars and critics as well as poets and novelists—writing, and the actual, even if only occasional, fulfillment of that hope is what keeps much textual production, literary and other, going. It is a practice that we all have a stake in preserving.*

---

* "What Was 'Close Reading'?: A Century of Method in Literary Studies" originally appeared in *The Minnesota Review*, vol. 8,7 pp. 57-75. © 2016, Virginia Tech. All rights reserved. Republished by permission of the copyright holder and the present publisher, Duke University Press. www.dukeupress.edu/

Chapter Six

Scientizing the Humanities: Shifts, Collisions, Negotiations

The title of this essay can be interpreted both narrowly and broadly—and, of course, as either appreciative or critical. It refers most broadly, and more or less appreciatively, to efforts on the part of scholars in humanities disciplines to introduce concepts, methods, or findings from the natural sciences into their home fields, usually in order to illuminate the customary objects of study in those fields: texts and artworks; writers and artists; ideas, human practices, historical events; and so forth. Such efforts are not new but, over the past decade or two, have become considerably more extensive, more programmatic, and more self-consciously science-allied than ever before. "Scientizing the humanities" also refers, more narrowly and rather more skeptically, to efforts that seek, as it is said, to "integrate" one or another humanities field with one or another science. Such efforts are reflected in calls for ongoing collaborations between scholars and scientists in particular fields (for example, between art historians and neuroscientists) and in the growing prominence of hybrid fields or approaches such as neuroaesthetics, literary Darwinism, or cognitive cultural studies. Thus, where literary scholars in the past might have explored Darwin's influence on the late Victorian novel or Gertrude Stein's interest in experimental psychology, they tell now of mammalian mating practices in *Pride and Prejudice*, the triggering of altruistic-punishment mechanisms in *Oliver Twist*, or the teasing of theory-of-mind modules in *Mrs. Dalloway*. At their most visionary, proponents of these approaches call for the total and terminal absorption of the humanities into the natural sciences, sometimes with rather millenarian-sounding promises and predictions.

The digital humanities are clearly a related development. Here efforts are not so much to make the humanities more scientific (though that is often an element) as to attune their practices more closely to the increasing power and presence of information technologies. Again, though such efforts are not new, they are considerably more extensive and programmatic than ever before. Where, in the past, a professor of English might have demonstrated the advantages of computer use to a principled Luddite down the corridor, now groups of scholars in literary "laboratories" across North America build software platforms and access, count, chart, and correlate huge databases of digitized materials to various ends, some more consequential than others. One cannot be simply "for" or "against" these developments. What I shall do here is indicate some considerations—historical, conceptual, and pragmatic—that I think are useful for understanding and assessing them.

I

The term that I have been using, *developments,* may appear too tame to many involved in the new approaches. Apocalyptic announcements abound and draw on a general sense, especially among premillennial academics, that a giant hinge has turned in the past decade or so, that our worlds—our students, the university, the culture, our own everyday practices—have, for better or worse, changed radically. The term *revolution* is not the one commonly used, but talk of seismic or tectonic shifts is pervasive. Among the reasons given for the announced upheavals, two are especially prominent. One is the ubiquity of information technologies and their rapidly growing centrality in our lives. The other is the deluge of what is claimed as radically illuminating news about ourselves issuing from the biological and behavioral sciences: news, especially, about our genes, our brains, and our evolutionary histories. I turn below to how this sense of a fundamental shift plays out in the scientizing projects I have been describing, but first I want to say something about the views of intellectual history implied in the discourses that promote them.

There are a number of models of the dynamics of intellectual history, models, that is, of how ideas and related practices, including scientific ones, change. Three of them are especially relevant here. One, a familiar model, theological in origin but associated with popular ideas of science, is of a general progressive movement from darkness to light: onward and upward, from ignorance and error to knowledge and truth.

This is the model favored by scientizers inspired by the idea of "consilience" developed by the sociobiologist E. O. Wilson. In his 1998 book *Consilience: The Unity of Knowledge*, Wilson represents Western intellectual history as a set of increasingly enlightened efforts, moving steadily toward harmony and unity since the seventeenth century, with disruptions from two major counter-enlightenment forces: Romanticism and postmodernism, as he names them. Attached to this model of history is the idea of an intrinsically hierarchical organization of knowledge—a chain or ladder of explanatory authority, with physics seen as foundational to all other scientific pursuits and biology seen as foundational to both the humanities and the social sciences. In accordance with this view, the classic Western project of enlightenment will be consummated when the humanities and the social sciences (seen as now carelessly scattered and willfully isolated) join that progression toward harmony and unity so that the destined integration of all knowledge, from bottom to top, can be completed. Thus Wilson writes in *Consilience:* "When we have unified enough certain knowledge, we will understand who we are and why we are here" (Wilson 1998, 1). As Wilson himself acknowledges, this vision amounts to a naturalized millenarianism.[69] The views of science on which it is based have been seriously challenged—many would say rendered obsolete—by a century of empirical work in the history of science.

A second, more historically informed view of the dynamics of scientific change is associated with the work of Thomas Kuhn, especially *The Structure of Scientific Revolutions*, first published in 1962. Talk of "paradigm shifts" and "epistemic breaks" by promoters of the new scientizing approaches draws implicitly on this second model, although, being promotional, they tend to retain major elements of the onward-and-upward story that Kuhn sought to displace.[70] A third model of the dynamics of intellectual history originates in Ludwik Fleck's *Genesis and Development of a Scientific Fact* (1979 [1935]). Like Kuhn's account of science (which it influenced), Fleck's account challenges the familiar progressivist story. But it is more sociologically acute, more responsive to cultural history, and also more radical with regard to ideas of knowledge and truth. In a word, it is constructivist rather than realist.

In Fleck's model, intellectual history is a dynamic field made up of the activities of multiple, distinct "thought collectives," that is, groups of intellectually interacting people, such as the members of particular religious sects or particular academic disciplines, and the ideas, discourses, and practices that they share. Scientific disciplines and academic fields of study are, in this model, neither hierarchically organized nor fixed in form; rather, they are continuously forming and transforming, sometimes merging and sometimes attenuating. Although the activities of fields and disciplines do not progress toward any general destiny, either unity or truth, they do issue in significant local achievements, including more or less radical conceptual innovations with relatively stable, broadly appropriated practical applications. There is much to be said for Fleck's views, and I have said more about them here and elsewhere.[71] They are of interest in the present context because they offer a well-developed alternative to the empirically dubious model of intellectual history that underwrites Wilson's program of pandisciplinary consilience and related calls for integrating the humanities with the sciences.

II

A fundamental consideration in assessing the new scientizing approaches is their relation to the aims and perspectives of the humanities as distinct from those of the natural sciences and, in the case of the digital humanities, as distinct from those of computer engineering. A recent article by Katherine Hayles is useful in highlighting the issues. Hayles is a longtime advocate of connections between the sciences and humanities, an influential analyst of all things digital, and a major proponent of posthumanism (or, at least, one of the sets of theoretical perspectives so-named). The article with which I am concerned is titled "Cognition Everywhere: The Rise of the Cognitive Nonconscious and the Costs of Consciousness" (2014). Readers familiar with the digital humanities scene will recognize an allusion to the idea of ubiquitous computation or, in a phrase used by one of its advocates, "computation everywhere" (Wolfram 2014). Part of Hayles's effort in her article is to suggest a comparable ubiquity to cognition, which, in her view, is properly understood to include the information-processing activities of mechanical as well as biological systems. Her major aim in the article, however, is to counter arguments by influential literary scholars to the effect that projects in the digital humanities fail to satisfy certain important disciplinary interests addressed by more traditional methods of study.

"Many print-based scholars," Hayles writes, "see algorithmic analyses as rivals to how literary analysis has traditionally been performed, arguing that digital-humanities algorithms are nothing more than glorified calculating machines" (213). Such objections, she believes, are based on scholars' ignorance of the current capacities of computers, along with an exaggerated sense of the importance of consciousness and of the distinctiveness or worthiness of human cognitive capacities more generally. Accordingly, she devotes much of the article to describing the human-like things that computers can now do—for example, "learn languages," "draw inferences," "compose music"—and, conversely, to detailing the limits and frailties of human cognition and consciousness as revealed by neuroscience and compared with the operations of computers. Thus she points out that computers used in financial markets can now process information automatically at speeds measured in millisecond differences, thereby providing enormous advantages to the traders using them, and that comparable advantages can now be obtained in the humanities, where computers, operating without the "presuppositions or biases" that come with human cognition and consciousness, "allow questions to be posed that simply could not have been asked or answered using human cognition alone" (212).

Hayles does not explain why literary scholars should want to pose and answer questions—presumably about works of literature, individually or in sets of various kinds—that they would not have asked or could not have answered using their own human cognitive capacities. The reason commonly supplied by advocates of the digital humanities (for example, Moretti 2000 and Jockers 2013, as discussed in chapter 5) is that *knowledge*—or, with emphasis, "real," "objective," "factual" knowledge—is thereby increased. But the explanation raises a number of other questions: What aims or interests are served by a sheer increase of factual information about some thing or set of things? Does a mere increase of objective, factual information constitute what we usually mean by knowledge? And does a mere increase of objective, factual information about various of its objects of study—without connection to any interests or purposes—make sense as a project for any humanities discipline as such?

A proper appreciation of major advances in information technology and neuroscience, Hayles writes, "requires a shift in conceptual frameworks so extensive that it might as well be called an epistemic break" (218). "Today," she cautions, "the humanities stand at a crossroad." One path "reinforces the idea that humans are special, that they are the source of almost all the cognition on the planet, and that human viewpoints therefore count the most in determining what the world means." On the other, better, path, scholars in the humanities would accept an enlarged "idea of cognition to include [the] nonconscious activities" of technical devices as well as of other biological systems (216-17). "With the resulting shifts of perspective," she believes, "many of the misunderstandings about the kinds of interventions the digital humanities are now making in the humanities [would] simply fade away" (218).

In seeking to correct what she takes to be misunderstandings of the digital humanities on the part of humanities scholars, Hayles strives to be informative and conciliatory. But her major efforts, I think, miss the point of many critics' concerns. Noting the growing significance of what she names here "the cognitive nonconscious," she writes: "One conclusion seems inescapable: the humanities cannot continue to take the quest for meaning as an unquestioned premise for their ways of doing business" (199). The phrasing suggests some misunderstandings on Hayles's part. Humanities scholars do not generally see a (or "the") quest for, or provision of, meaning as a central goal or premise of their activities. Nor do critics generally see the inability of computers to come up with the meaning(s) of texts (or of anything else) as the crucial limit of digital humanities. It's not that computers cannot produce or "interpret" textual "meanings" in some senses of those terms; they already can or soon will. The problem is that many of the algorithmic performances and productions currently invoked as examples of the achievements or promise of computers lack the type of interest that we find in the performances and productions of fellow humans as such. Calling the computational activities of technical devices "cognitive" and noting their similarity to actions performed nonconsciously by humans does not erase the sense of a crucial difference between the two or supply the type of interest—attraction, concern, connection, fascination, delight—found specifically in the latter.[72] Contrary to charges commonly leveled by enthusiasts of artificial intelligence, artificial life, and other computational wonders, the interest in question does not reveal a prejudice in favor of carbon- versus silicon-based "cognition," "intelligence," or "life." What makes the actions, performances, and productions of other humans—writers and composers, artists and critics, kings and revolutionaries—especially interesting to us is not our conviction that humans are superior to machines and nonhuman animals; it is our recognition that they are the same sorts of machines and animals that we ourselves are.

A good part of the interest of the actions and productions of other humans may have to do with our experiencing the world, fairly uniquely among machines and animals, as subjects—experiencing it, that is, with what we call consciousness or a sense of self. Hayles, having perhaps heard such observations from digital-resistant humanists, goes to some trouble to expose subjectivity, consciousness, and a sense of self as illusions. But the effort is, again I think, misplaced. Recognizing that subjective experiences—one's own and other peoples'—are, as she terms them, "epiphenomena of underlying material processes" (202-203) does not make them any less interesting as experiences. Nor does it erase the difference that we generally register—perceptually, conceptually, and emotionally—between experiencing beings as such and material processes as such.

Hayles writes of the anthropocentric bias that attends the operations of consciousness in humans, a result, she explains, of our (illusory) sense of possessing a particular self and our concern for its well-being. This bias, she suggests, leads humans to overestimate their importance in the world and their ability to control the complex ecological systems in which they are embedded, with various ecological catastrophes among the consequences. Aside from the suggestion of a proper estimate of humans' importance in the world (who could arrive at such an assessment, and how?), this observation is no doubt true. But the anthropocentrism of the humanities is as definitive as the astro-centrism of astronomy or the bio-centrism of biology, with the addition, among humanities scholars, of a type of interest in the defining objects of their study—that is, human ideas, artifacts, practices, and events—that comes from a particular bond of kinship with the authors, agents, and subjects of those ideas, artifacts, practices, and events.

This discipline-defining anthropocentrism does not require a particularly high regard for all things human. It certainly does not require a refusal to recognize our biological nature or cognitive limits. Hayles argues that the news from neuroscience and due recognition of the ubiquity of the cognitive nonconscious together undercut standard views of human rationality and the power of reason. Outside some departments of philosophy, however, it is generally not scholars in the humanities who overvalue rationality. After all, the idea of reason has not had a very good press among writers, critics, and theorists for some time now—one may think of the doubts about it raised (as E. O. Wilson is aware) by the Romantics or of its treatment by Nietzsche or psychoanalytic theory. Nor is it humanists who need to recognize the existence of what Hayles calls "systemic human blindnesses." On the contrary, if we have a concept like hubris and a chastened sense of human capacities more generally, it has come largely from poets, humanistic philosophers, and those who study and transmit their views. Humanities scholars these days generally acknowledge—and many of them stress—the continuities between humans and other animals; and, although a strong suspicion of a not well-understood Darwinism remains widespread, most of them, I believe, would acknowledge that our capacities, impulses, and responses reflect, among other things, the evolutionary history of the species. Scholars in the humanities may be inclined to add that the capacities, impulses, and responses of humans also reflect our relatively complex neural organization and are shaped by the evidently unique existence, among us, of language and intergenerationally transmitted artifacts, ideas, practices, and institutionalized norms. But most evolutionary biologists and neuroscientists would be inclined to note the same things.

III

Hayles's evocation of a decisive crossroads for the humanities recalls a comparable evocation by literary Darwinist, Joseph Carroll. In an article titled "Three Scenarios for Literary Darwinism" (2010), Carroll describes three possible future trajectories for the critical approach that he founded and promotes. In the first scenario, literary Darwinism would remain a minor movement. In a second more hopeful one, the movement would become mainstream but still only as one among other "'approaches' to literature." In the third scenario, which Carroll urges, the field of literary studies, along with all other humanities fields, would be totally transformed by evolutionary theory and integrated with anthropology, economics, sociology, and political science—all similarly transformed—to make up a new field that he calls "the evolutionary human sciences."

Those familiar with literary Darwinism will recall that scholars pursuing this approach seek to explain why we read poems or novels, and also why authors write them and, sometimes, why fictional characters behave as they do, in the same way that evolutionary psychologists explain virtually everything else that humans do—that is, as manifestations of the operation of putative universal, hard-wired mechanisms that evolved to enhance the reproductive fitness of our Stone Age ancestors. My concern here is not with the assumptions, methods, or claims of evolutionary psychology (I have examined them elsewhere)[73] but with the idea, promoted by Carroll and other literary Darwinists, that those assumptions, methods, and claims should be the foundation of literary studies and, in Carroll's case, of all other humanities fields as well.[74]

Toward the end of his article, Carroll observes that the future of literary Darwinism is hard to predict. If his third, integrationist scenario fails to be taken up by humanities scholars, it will be, he writes, because of an entrenched "mind/body dualism" and an ideological "pluralism" based on habit, convention, and scholars' ignorance of science. On the other hand, he continues, if literary studies joins the other evolutionary human sciences, then "the institutional resistance of the postmodern establishment will crumble from within … as a result of intellectual dry rot" (60) and a rich and pleasant prospect will open for those who remain:

> Aspiring literary scholars will have open before them a wide spectrum of methodological choices, ranging from the purely discursive, essayistic form of commentary that now dominates the humanities to the rigorously quantitative, empirical methods that now prevail in the sciences …. [Graduate students] will not cast about desperately for novelty, taking recourse in superficial verbal variations ensconced in sophistical theoretical ambiguities. They will, rather, wake up like kids

at Christmas, delighted with the endless opportunities for real, legitimate discovery that are open to them. (64)

He concludes with the evidently non-ironic observation that the third scenario will be hard to achieve but that "it promises discovery, things not yet dreamed of, lying in the bosom of reality" (64).

Carroll has little good to say of literary studies as traditionally or, especially, as currently pursued. Hayles's evocation of forking paths for the humanities has none of Carroll's biliousness, but the alternatives she offers are, in some respects, similarly framed. Like Carroll, she finds literary studies crucially deficient and, like him, she is optimistic with regard to the future if the avenue she urges is followed. In Hayles's account, while the cost of taking the wrong path is the "isolation of the humanities from the sciences and engineering," with the right path, "the search for meaning [duly understood as "information flows"] then becomes a pervasive activity among humans, animals, and technical devices, with many different kinds of agents contributing to the rich ecology of collaborating, reinforcing, contesting and conflicting interpretations" (218). Some features of the traditional humanities, Hayles writes, will have a place amid this interpretive multiplicity and diversity: "The sophisticated methods the humanities have developed for comparing different interpretations then pay rich dividends for other fields and open up to any number of exciting collaborative projects" (218). She concludes encouragingly: "The humanities can make important contributions to such fields as architecture, electrical and mechanical engineering, computer science, industrial design, and many other fields" (217-18).

I suspect that the prospect of exciting collaborations with mechanical engineers and computer scientists would not persuade many premillennial humanities scholars and teachers to give up their privileging of the study of ideas, artifacts, and individual texts by individual human beings. It might well be attractive, however, to a good number of young researchers already at home in the world of information technology—blogs and games, platforms and programs—and especially to those already engaged in computational projects. Similarly, Carroll's promise of endless opportunities, via evolutionary psychology, for "real, legitimate discovery" "in the bosom of reality" may appeal to some graduate students of literature already captivated by a certain idea of science and of what it is to be genuinely scientific. But it is not likely to end the attraction of a good many of them to "superficial verbal variations ensconced in sophistical theoretical ambiguities"—or to what Carroll hears as such. I return below to the significance of these differences of disciplinary training, intellectual taste, and personal temperament.

IV

I would like to turn directly now to what could be described as the historically distinctive aims and achievements of the humanities. They are not easy to describe and current celebrations of the humanities tend to be nostalgic and selective at best and, at worst, vacuous.[75] According to Carroll and others (Dennett 2006, Slingerland 2008), resistance to the new scientizing approaches reflects an obsolete mind/body dualism that places an artificial barrier between the humanities and the sciences. The charge is misdirected but useful for sharpening the terms of the differences at issue. There is, of course, a longstanding theologically grounded insistence on the distinctness of the realms of the spiritual and the material; and it is often allied with the claim that there are forms of knowledge—revealed or intuitive, for example—that are higher or deeper than scientific findings or with the idea that there are phenomena, such as consciousness or "products of the human mind" that cannot be explained in physical terms or approached naturalistically. But resistance to the idea of integrating the humanities and the natural sciences does not require any of those hoary dualisms. It is not that the disciplines are properly attached to distinct realms of being but that humans orient themselves to the phenomenal world in multiple ways and that these orientations are reflected in the different aims and practices of the various arts and sciences and, in the West, of the various academic disciplines.

One of the ways we orient ourselves toward the world is by seeking to extend our knowledge of, and strengthen our control over, the physical conditions of our existence. Accordingly, we seek to chart, model, and explain those conditions conceptually and to modify them or intervene in their operations technically. But humans everywhere also seek to develop and manifest themselves as experiencing creatures and, accordingly, are commonly engaged by the experiences, creations, and reflections of their fellow humans. These two are not the only ways that we orient ourselves toward the world; the list could be extended. But they are clearly distinct, and they evoke the different aims and practices that, in the West, have become specialized, or roughly specialized, as the natural sciences and the humanities.

Contrary to the suggestion and conviction of many promoters of Wilson's "consilience," the specific value of the modern sciences does not lie in their ability to deliver "certain knowledge."[76] It lies, rather, in what has evolved historically as a set of conceptual commitments and related practices attached to aims that are, variously, both pragmatic and intellectual. The commitments in question, notably naturalism, empiricism, and experimentalism, along with the types of practice they entail, constitute an extremely efficient apparatus for generating models of the operations of the physical world that enable us to predict, control, and intervene in those operations effectively and reliably. To the extent that any human project has those sorts of aims, the apparatus of modern science is probably the most consistently effective means for achieving them.

Scientific theories, models, and accounts also, but less centrally, respond to our desire for intellectually satisfying explanations and interpretations of the phenomenal world. They are less central in this regard because the experience of intellectual satisfaction is considerably more variable than the observation of pragmatic effectiveness. All may agree that a bridge has been built, and most may agree that an ailing baby has been cured. But an explanation of some complex and humanly significant set of phenomena—say, of art, love, or religion—that some people find uniquely adequate may strike others as superficial and still others as clearly improper. The more consistent reliability of empirical, experimental, and naturalistic models and explanations in serving pragmatic aims does not make them the only kinds of accounts of the phenomenal world that we value or the only ones generally recognized as knowledge.[77]

I have referred to certain aims and achievements of the humanities as historically distinctive. The word *historically* is crucial here. The humanities are not an essential or natural kind. They are clusters of contingently institutionalized custodial, intellectual, and pedagogic practices. For the past four hundred years or so, those practices, as pursued by Western and Western-educated scholars, have included the identification, preservation, description, analysis, explication, dissemination, and often—but not always—celebration of what are regarded, at any given time, as significant human events and cultural achievements.[78] "Research" in the humanities—or, as we say, "scholarship"—is commonly understood as the disciplined pursuit of such practices. "Study" in the humanities is commonly understood as the acquisition of expert knowledge of some specific body of materials (largely but not exclusively textual) and the development of techniques and skills—for example, archival, philological, musicological, iconological, or analytic—required for the pursuit of such practices. This description is not, I think, notably tendentious. But it does suggest the *distinctive* intellectual character of the humanities disciplines and their *specific* institutional, social, and cultural functions.

The argument made by Wilson, Carroll, and others promoting the new scientizing approaches is not that study in the humanities should be more closely engaged with or better informed by the natural sciences. Moves in these directions are, in my view, long overdue and, where they occur, to be applauded. The promotional argument is, rather, that the study of art, literature, music, philosophy, and so forth should be more "scientific" in method and aim, with desirable method usually described as "quantitative" and "objective" and desirable aim usually referred to—in pointed contrast with whatever the humanities are thought to seek or achieve—as "serious," "genuine," or (especially oddly) "adult" knowledge.[79]

It is true that our aesthetic, critical, and reflective engagements with the world do not produce what is usually thought of as scientific knowledge. But engagements of those kinds do have significant effects, including intellectual ones. Their effects are not as palpable, demonstrable, immediate, or pragmatically translatable as the products of our investigative or interventionist engagements with the world, but they can be consequential. To the extent that the pedagogic and scholarly practices of the humanities elicit, enable, and shape such engagements, they also can be, in the same way, personally and communally consequential: important for our continued development, both individually and generationally, as responsive, creative, critical, and reflective creatures.

Traditionally in the humanities, one "studies" the phenomena of art, literature, religion, and philosophy—that is, human creations, practices, and ideas—in the sense of examining them closely: not, however, usually just to gather facts about them or just to register their empirically describable features but, rather, with a view to understanding and elucidating the motives and experiences involved in their production and reception. Exploring, describing, and seeking to understand motives and experiences are fundamentally different from counting, measuring, and seeking to explain empirically observable phenomena. *Seeking to understand and convey experiences* is fundamentally different from *seeking to explain behaviors*. The humanities are, in that respect, typically first-person or, in a term from anthropology that I prefer, *emic*, that is, operating from the perspective of participating insiders, rather than third-person or, in the corresponding term, *etic*, that is, operating from the perspective of observing outsiders. Typically *partisan* rather than impartial, the humanities are an institutional locus not of disinterested interest in humans as one biological species among others but of distinctly self-interested concern for species capable of conscious experience—which, as we know from our interest in animal fables, cartoons, and science fiction, exceeds the species *Homo sapiens* by quite a bit.

It is certainly possible to study human practices, beliefs, and cultural products as "natural phenomena." For example, paintings, poems, and philosophical essays can be compared to the material products or bodily displays of other creatures, such as anthills and peacocks' tails. Comparisons of this sort are not unusual in the work of sociobiologists and, following them, literary Darwinists.[80] And, of course, human practices and their various products and traces can be investigated and described in strictly quantitative, physical terms without reference to individual human experiences. Doing so is standard practice in fields such as demography or economics and, following them, in the digital humanities.[81] Investigations and descriptions of these kinds undoubtedly produce facts about certain aspects of the usual objects of humanistic study. The question commonly raised by those resisting the new scientizing and computational methods is to what extent the production and possession of such facts serve the traditional aims of the humanities.

Since I don't think the answer to that question is obvious, I want to comment briefly on disciplinary aims and methods. In humanities scholarship, as in any domain of human activity from agriculture to deep-sea diving, methods are usually developed to further existing purposes. Here as elsewhere, however, the relation between method and purpose can be complex. Humans are curious, manipulative, and inventive creatures, and our purposes are continuously enlarged and transformed by the availability of new methods and their associated instruments. Whatever the initial purpose for which an instrument was fashioned—whether stick, bowl, or computer—we are likely to discover other useful or interesting things we can do with it; and those novel activities will generate new purposes and instruments, which will require and privilege new skills and talents. For these reasons, the invocation of past purposes and current practices as the sole criteria for assessing new methods in the humanities amounts to a stultifying conservatism. But resistance to various scientizing or digitizing methods may have a more substantive component, namely, that their outcomes—the evolutionary explanations, the cognitive redescriptions, the computer-generated correlations—appear crude, banal, or trivial by virtually any measure of intellectual value. To the extent that advocates of the new methods ignore such criticism, they produce their own self-immurement and stultification.

The differences I've been noting in the epistemic orientations and social functions of the humanities and the natural sciences are significant but, as I have emphasized, they are historical, not intrinsic. Nothing holds disciplinary differences in place but ongoing practices and their relation to the broader social collective. Insofar as distinctive practices and social functions exist, however, they have important practical implications for the new hybrid programs.

V

Academic and scientific subfields—for example, eighteenth-century French literature, high-energy particle physics, or Lacanian film studies—are what Ludwik Fleck called "thought collectives." They are distinguished from one another not only by subject domain or what aspect of the phenomenal world they study but also, and no less crucially, by implicit systems of linked assumptions, discourses, and technical practices or what Fleck called "thought styles." Kuhn called them, or something like them, "paradigms." I describe them elsewhere (Smith 2005/6,

108-129) as "disciplinary cultures." The awkwardness of the disciplinary newcomer—the art historian bumbling in the laboratory of the neuroscientist, or the neuroscientist bumbling in the gallery of the art historian—is like that of any other cultural immigrant. Becoming acculturated as a neuroscientist or an art historian is not just a matter of mastering a set of canonical ideas, texts, images, or techniques. It is also a matter of knowing a set of tacit but crucial norms: what counts as a well-designed experiment, a useful model, a rigorous analysis, or a subtle interpretation. It is a matter of knowing what matters: the important issues in the field, the significant rival views, which connections are crucial and which irrelevant. It is having a fund of informal know-how acquired through a personal history of active practice as, precisely, an active practitioner.

The existence of highly specific disciplinary cultures creates difficulties for any interdisciplinary venture, even when the conjunction involves closely related but historically distinct fields, such as evolutionary and developmental biology, now joined in the field of "evo-devo." Thought styles are powerful in shaping perceptions as well as discourses and practices, and there can be chasms of mutual incomprehension between members of different thought collectives, including academic disciplines and the subfields within them. One may think here of clinical and experimental psychology or of analytic and continental philosophy. The difficulties increase, of course, as the fields involved are more diverse in aim and orientation, and some difficulties are specific to ventures seeking to merge humanities fields with natural sciences or engineering.

An especially significant set of problems arises from the long-standing prestige differentials among academic disciplines, which exactly mirror Wilson's hierarchy but in reverse: here, physics is at the top while fields such as art history or literary studies are at the bottom. (Although there are no intrinsic hierarchies among disciplines, there are, of course, de facto dominances.) Consilient engagements between humanities scholars and scientists or engineers would presumably work well in both directions. That is the hope, claim, and promise of the new collaborative ventures. But the prestige differentials here are very steep, and the forces sustaining them draw on other invidious distinctions in the culture of the academy and in the broader culture as well. They are reflected in familiar contrasts between hard and soft disciplines, between real things and mere words, and, as noted above, between serious work and mere play.

Prestige differentials are significant in this context because they exacerbate a number of perennial problems in projects that seek to cross the Two Cultures. One is the tendency of humanities scholars to regard the scientific and technical materials that they import—findings, concepts, and methods—altogether uncritically, even when, as is largely the case in the new hybrid fields, those materials are still being developed and are still controversial in their own scientific disciplines. Duly studious humanities scholars may become quite knowledgeable about findings, concepts, and methods in fields such as evolutionary psychology or cognitive neuroscience, and they can be quite adept at summarizing them for fellow humanists. But they are generally not equipped to assess experimental designs, statistical analyses, or the robustness of conclusions in those fields. These evaluative skills come with training and experience working in the fields themselves, which, as I have already noted, bring practicing scientists detailed knowledge of current theoretical and methodological issues and important rival approaches. Because scientizing humanities scholars are at a disadvantage in these respects, they often put their money on transient ideas and methods. For example, assumptions about human development and evolution central to evolutionary psychology and literary Darwinism are questioned by established scientists and theorists in related scientific fields, such as genetics, evolutionary biology, and developmental psychology (see, for example, Oyama 2000 and Pigliucci and Müller 2010). Similarly, concepts in the neurosciences that figure prominently in cognitive approaches in the humanities—from the significance of mirror neurons for human behavior to the existence of a specific theory-of-mind module—are undergoing extensive modification in those fields (see, for example, Cook et al 2014).

No less significant for the temper of would-be consilient collaborations are the sorts of differences mentioned above in regard to the likely uptakes of Hayles's and Carroll's hopeful visions. There are not, in my view, two different types of people in the world, humanities types and science types. But in project-related interactions between scholars and scientists or scholars and engineers, strong differences of personal and intellectual temperament as well as talent, taste, and style are likely to give rise to severe cognitive dissonances in both directions.[82]

VI

Many of the difficulties traced here have been noted by scholars and scientists who themselves work in hybrid fields. For example, Johanna Drucker, a historian of graphic design and a major theorist of the digital humanities, calls attention to a fundamental clash between, on the one hand, qualities such as complexity, ambiguity, and indeterminacy that are generally appreciated in the humanities and, on the other hand, qualities such as simplicity, clarity, and predictability that are highly valued in computer engineering and reflected in the binary character of information technology itself (Drucker 2012). In its "rush to be computational," Drucker suggests, digital scholarship is in danger of forgetting hard-won theoretical perspectives in the humanities, among them constructivism and relativism, which she names explicitly.

Similarly, Anjan Chatterjee, a neuroscientist who also conducts research in the hybrid field of neuroaesthetics, writes of the fundamental challenges faced by efforts to bring brain science into the humanities fields traditionally concerned with aesthetics: art history, literary theory, philosophy, and so forth. Chatterjee asks, "When does neuroscience provide deeper descriptive texture to our knowledge of aesthetics, and when does it deliver added explanatory force?," and comments:

> Knowing that the pleasure of viewing a beautiful painting is correlated with activity within the orbito-frontal cortex ... adds biologic texture to our understanding of the rewards of aesthetic experiences. However, it is not obvious that it ... advances our understanding of the psychological nature of that reward. For neuroscience to make important contributions to aesthetics, the possibility of an inner psychophysics has to be taken seriously. (Chatterjee 2011, 60).

The comment is urbane and perceptive. A correlation between someone's report of pleasure in viewing a painting and an image of activity in some area of that person's brain does not *explain* the pleasure. But it does add "biologic[al] texture" to our understanding of that kind of experience—a nice turn of phrase that reflects the contribution of a certain type of scientific knowledge to a classically humanistic enterprise while at the same time acknowledging its limits.

Connecting an observation of neuronal activity to a reported experience of pleasure is partly a conceptual problem: the classical philosophical conundrum of mediating between third-person observations and first-person experiences, which I have referred to as, respectively, *etic* and *emic* perspectives. But making such connections is also a rhetorical problem, a matter of finding some way to articulate—join together—two verbal-intellectual idioms that have evolved historically to serve significantly different ends: on the one hand, the observational, impersonal idiom of the natural sciences, which strives to be informative and appropriately precise; and, on the other hand, the phenomenological, experiential idiom of the humanities, which strives to be evocative and appropriately subtle. Negotiating these two perspectives and joining these two idioms is not an impossible task.[83] But it requires a kind of intellectual-linguistic tact, the cultivation of which is one of the major challenges faced by the new hybrid approaches.

VII

What we speak of now as "the natural sciences" and "the humanities" are only relatively stable assemblages of continuously emerging, developing, combining, and differentiating intellectual traditions and practices. Neither is likely to retain its current forms or even its identity in the future. On the contrary, we are witnessing major transformations and attenuations of both in our lifetimes. The "shifts"—though not quite "tectonic"—are real enough. As the humanities become increasingly scientized, the sciences themselves are becoming increasingly industrialized and commercialized (Shapin 2008, Mirowski 2011). Intellectual historians already have reason to ask, "What was 'science'?" Sooner or later they will have reason to ask, "What was 'classics'? What was 'art history'?" and—perhaps especially puzzling—"What in the world was 'English'?"

There is little reason to think the humanities will fold themselves into the natural sciences and, I believe, no good reason to think they should. But there *are* reasons to think that at least some of the new hybrid approaches will survive and prosper. For one thing, they are attracting a good number of talented, energetic, and broadly informed young people. For another, considerable institutional resources are already invested in them. Significantly, practitioners have begun to respond to external criticism constructively rather than with defensive hostility and also to engage in discriminating internal criticism rather than indiscriminate mutual puffing.[84] As increasingly mellowed, chastened, and sophisticated products of the new approaches—neuroaesthetics, cognitive cultural studies, digital humanities, and so forth—appear in journals, classrooms, and conferences, they will begin to join other practices in the humanities academy, both old and new, both disciplinary and interdisciplinary.

In Fleck's model of intellectual history, disciplinary transformations, though continuous and sometimes radical, are not always revolutionary and never total. As new approaches make headway, established ones commonly continue for some time, typically transformed—sooner or later—by the most significant new methods and ideas. At Duke University in the 1990s, even as its English department became notorious for ideology critique, reader-response criticism, and queer theory, members of its faculty were teaching texts in Middle English and preparing editions of William Faulkner's novels and Thomas Carlyle's correspondence. Texts are still taught there in Middle English, although now with quite different emphases. The specific activities I have described as historically attached to the humanities disciplines—namely, identifying, preserving, elucidating, and disseminating significant cultural achievements and the record of significant human events—are not the only things that humanities scholars do. But they are activities that are valuable for the human collective at large and that humanities scholars do more or less uniquely and more or less effectively. If academic scholars no longer do them, one must hope they are done by other agencies, human or nonhuman, formally or informally. I don't think the historically distinctive activities of the humanities disciplines will disappear. But they may be dispersed: not housed in a distinct quarter of the academic world and perhaps not housed in the academy at all. Many of those activities were performed in the past outside the academy—for example, domestically, or by clergy or hired tutors. Many have already been transferred largely to electronic venues—for example, to online courses, blogs, and websites—and one may anticipate further dispersals to venues of that kind.

There is much in what I have described here to give us pause and perhaps to make us weep. Two further considerations, however, can be heartening. First, there is good reason to think that, even with the attenuation of "print culture" and the flat-out disappearance of "classics," "English," and even "philosophy," humans across the globe will still be inclined to recall, savor, and ponder what fellow humans have done, made, and articulated, no matter how—or via what medium—it is transmitted. Second, although desegregations and new mixtures typically elicit fears of a homogenized or mongrelized future, both cultural and biological history remind us that hybrids often turn out to be sturdier than their ancestors and, indeed, to be especially favored in surprising ways. The traditional Western disciplines, both the sciences and the humanities, are being severely shaken up by important intellectual and technological developments. But the disciplines—again, all of them—are also being put together in myriad new ways. The new disciplinary configurations are not, in my view, moving toward ultimate harmony or unity. But they may be opening out to intellectual landscapes more interesting than most of us imagine.*

---

* "Scientizing the Humanities: Shifts, Collisions, Negotiations," originally appeared in *Common Knowledge*, vol. 22:3, pp. 353-372. © 2016, Duke University Press. All rights reserved. Republished by permission of the copyright holder. www.dukeupress.edu

Chapter 7

Perplexing Realities: Practicing Relativism in the Anthropocene

This chapter began as a talk prepared for a colloquium titled "Climate Realism." Its conveners asked, "How is realism—in both the aesthetic history of representation and the philosophical tradition that underwrites it—transformed by contending with our new experience of climate in the Anthropocene?"[85] For many people, ideas of reality, in the sense of taken-for-granted beliefs about what the world is like, have been duly updated, radically unsettled, or otherwise affected by reports of climate change. In a good many cases, such ideas have been affected directly by experiences of the forces and effects involved. Aside, however, from various renewed invocations and amplified self-endorsements, the realist tradition in philosophy does not seem to have been transformed by those reports or such experiences. Nor has work in that tradition, either classic or current, been notably significant in illuminating either of them. With regard to the perplexities of climate change as elsewhere, however, the traditions of constructivist thought described in previous chapters offer useful perspectives for conceptualization and practice.

Consideration of three such perplexities will illustrate the point. One is how to respond to malign appropriations of constructivist critiques of standard accounts of scientific knowledge. An essay by Bruno Latour often cited in that connection, "Why Has Critique Run Out of Steam?" (2004), provides a point of entry here. A second is how to understand climate "denialism," the rejection of the descriptions, explanations, and predictions of climate science by large segments of the public. Here Ludwik Fleck's account of belief systems is illuminating, especially when joined with contemporary constructivist understandings of human cognition. A third perplexity is the existence of multiple, sometimes radically divergent, operative realities. Here the commitment of constructivist social scientists to methodological symmetry can be instructive for ethical-political practice as well as for conceptualization. At the end of the chapter, I reflect on a set of related problems, both conceptual and practical, in recent writings by environmentalists.

I

Realism, in academic philosophy, is the view that the objects we perceive and describe, with the features we observe and characterize, exist independent of our perceptions and ways of describing them—or, as it is often put, that there is "an objective reality" apart from what we think or say. This view is commonly attended by two other ideas. One is that a belief or statement is true insofar as it corresponds to that objective reality or represents the features of that world accurately. The other is that science is a distinctive method for discovering those features and for arriving at true beliefs about its workings. Thus philosopher of science Philip Kitcher, explaining the "broadly realist conception of science" he seeks to defend, describes its components (or, as he writes, its "old-fashioned virtues") as follows: "Scientists find out things about a world that is independent of human cognition; they advance true statements, use concepts that conform to natural divisions, [and] develop schemata that capture objective dependencies" (Kitcher 1995, 127).[86]

These views, formulated in more or less these terms, are familiar. They have, however, been subject to significant challenges. Skeptical philosophers since antiquity have pointed out that we cannot glimpse a putatively objective, autonomous reality around the corner of our own perceptions.[87] Of course, we (or the scientists among us) may chart the apparent regularities of our humanly shared perceptible world and, with various technologies, detect remote, complex, ordinarily imperceptible phenomena. And we (or the philosophers among us) may attempt to infer the essential, underlying features of such a putatively objective reality through, as it is said, the use of reason and logic. But what we (scientists and philosophers included) cannot do is conceive or describe the putative features of such a putative reality independent of any humanly derived concepts or discursive idioms.

Figure 1. A Glimpse of Reality? Anonymous woodcut in Camille Flammarion, *L'Atmosphère: Météorologie Populaire*, Paris 1888, p. 163. Source: Wikimedia Commons

Ontologies—purported accounts of what's really out there, or what's underneath it all, or what there really *is*—can be more or less conceptually congenial and pragmatically workable, and we may admire, endorse, or adopt various of them accordingly. What they cannot be, however, is "objectively correct." Contemporary philosophers have mounted important related challenges to conventional realist views.[88] The work of philosopher of science, P. Kyle Stanford, is of particular interest here. Arguing from the historical record, Stanford advocates an essentially pragmatist (he calls it "instrumentalist") view of the sciences and, with it, of our relation to the experienced world (Stanford 2006, 2016). Scientific theories are effective, he writes, not, as realists commonly maintain, "because they are true" but, rather, "because they help us successfully navigate the world in productive and systematic ways" (Stanford 2016, 96).[89]

In response to such challenges, academic philosophers have defended, qualified, and updated what they continue to name *realism*. Over the past few decades, for example, we have been offered "scientific realism," "realism with a human face," "post-positivist realism," and "speculative realism," each with a set of related ontologies and epistemologies, that is, conjectures about what really *is* and claims about how we (or some of us) can know it. Specific realist and neorealist positions are not my concern here, but I would make a general observation: with all due respect, climate change does not change *everything*.[90] Where conceptually problematic invocations of a knowable objective reality have been effectively challenged, their intellectual status is not redeemed by our appreciation of the evidence of global warming or by our frustration at the denial of that evidence. Likewise, important challenges to realist views of scientific authority are not made irrelevant by our appreciation of the efforts of climate scientists or by our awareness of campaigns to discredit those efforts.

The above observation evidently needs emphasis in view of persistent claims to the contrary: the claim, for example, that a robust philosophical realism is needed to counter the "alternative facts" spun out by political strategists, or that a "postmodern" critique of objectivism has led to a public mistrust of scientific knowledge.[91] Dubious claims of these kinds should, I think, be countered and rejected. But we may also acknowledge the intellectual and pragmatic perplexities they reflect regarding the realities of, among other things, climate change.

Some of those perplexities were given vivid expression by Bruno Latour in the essay mentioned above, "Why Has Critique Run Out of Steam?: From Matters of Fact to Matters of Concern." Published in 2004, it continues to be widely cited and remains influential in relation to these issues. In a passage of particular interest, Latour quotes a political strategist seeking to deflect concern about global warming who advises making "the lack of scientific certainty" a primary point in arguments. "Do you see why I am worried?" Latour asks. "I myself have spent some time in the past trying to show '*the lack of scientific certainty*' inherent in the construction of fact[s]" (227). He continues with a cascade of further questions:

> Was I foolishly mistaken? ... Was I wrong to participate in the invention of this field known as science studies? ...Why does it burn my tongue to say that global warming is a fact whether you like it or not? Why can't I simply say that the argument

> is closed for good? ... Should we apologize for having been wrong all along? Or should we rather bring the sword of criticism to criticism itself and do a bit of soul searching? ... Nothing guarantees, after all, that we should be right all the time ... Isn't this what criticism intended to say: that there is no sure ground anywhere? But what does it mean when this lack of sure ground is taken away from us by the worst possible fellows as an argument against the things we cherish? (227)

Here and elsewhere in the essay, Latour poses crucial questions and says powerful things. But, here as elsewhere, the ironies are complex, the allusions ambiguous, and the recommendations somewhat elusive or equivocal.

His aim, Latour writes, is to raise some issues about "critique": "not," he continues, "to depress the reader, but to press ahead, to redirect our meager capacities as fast as possible." But both the redirection proposed and the meaning of the contrast at the heart of it—from "matters of fact" to "matters of concern"—are open to widely differing interpretations. One interpretation, current in the literary academy, is that scholars should turn from critique to more positive activities: to *caring* about texts and other things; to taking writers at their word instead of "digging deeply" into their texts for hidden meanings; to being less concerned with facts and more with affects; and so forth. This interpretation is reflected in a recent article titled "The New Modesty in Literary Criticism." Its author, Jeffrey Williams, writes as follows:

> In the theory years, you *were* what your reading was—Marxist, feminist, deconstructionist, queer ... Some who were in that camp, with its suspicious habits, began moving away from it by the late 1990s and early 2000s ... Other scholars also began to worry about the effects of theory. In a much-discussed ... essay, "Why Has Critique Run Out of Steam?," the sociologist of science Bruno Latour worried that the relativism and social constructionism of postmodern theory had discredited good science. (Williams 2015)

That is not, of course, quite what Latour said was worrying him.

I suggest another way to interpret Latour's worries below but would offer a word, first, on my perspective on these issues. It derives in part from my association with the literary academy but is shaped more significantly by an early intellectual sympathy with the constructivist tradition in philosophy and the social sciences. The interest and sympathy were reflected in talks and articles I presented to academic audiences in the 1990s. At the time, the American academy was much occupied by the "science wars"—fought, it should be recalled, by those defending traditional, largely rationalist-realist views of science against those promoting the relatively novel, largely constructivist views emerging from science studies. I was myself much occupied at the time both expounding the latter views and attempting to untangle ignorant confusions and malicious conflations regarding them, for example, their identification as (or with) "deconstructionism," "postmodernism," or "social constructionism," and their conflation with Marxist ideology critique and/or, sometimes in the same breath, with classic idealism. There were lessons to be learned from those controversies about advocacy under conditions of extreme polarization and motivated distortion. These are, of course, just the conditions that we face today with regard to climate change and that Latour evokes in the opening pages of his essay. We may return, then, to what he probably was and was not saying there.

In urging a turn from "matters of fact" to "matters of concern," Latour was certainly not saying that we should revive triumphalist narratives of the history of science or realist-rationalist explanations of the authority of scientific knowledge. Nor was he repudiating either the critiques of such accounts elaborated over the twentieth century by constructivist theorists of science or the alternative accounts developed in the field of science studies by, among others, himself. Such critiques have continued to win his endorsement and the alternative accounts continue to figure centrally in his work, including his writings on climate change.[92] What he may have been doing in the essay, at least in part, was calling on his fellow researchers and theorists in science studies to turn from critical treatments of traditional epistemology (or "matters of fact"), which had lost their urgency ("run out of steam"), to more concrete and pressing problems (or "matters of concern"), such as controversies over climate change, where their expertise might be useful. Such calls for increased engagement with current problems were already familiar in the field and remain current in it, along with debates over scholars' commitment to neutrality versus explicit advocacy (Demeritt 2006, Zuiderent-Jerak and Jensen 2007). This interpretation lacks the drama and scope implied by the title of Latour's essay and by various allusions in the text.[93] It has the advantage, however, of making sense in relation to his work as a theorist and practitioner of science studies. It also allows us to have some grip both on the "worries" he evokes rather obliquely in that opening passage and on the responses to such worries that he seems to suggest there.

The fact that "the worst possible fellows" say things that sound like what we say certainly presents problems, but they are not new ones. Indeed, they are ancient and proverbial. The devil or assorted devils have always quoted scripture for their own purposes. That has never been thought a good reason to stop preaching, much less to abandon scripture. It *has* been thought a good reason to clarify one's teachings and to pay close attention to the situations and idioms of those one seeks to address. Constructivist accounts of science are, of course, not scriptural. As it happens, though, this is the solution to which Latour himself turns after detailing his corresponding worries, as a quasi-theologian, about questionable treatments of the New Testament. The title of the work in which he spells this out is *Rejoicing: Or the Torments of Religious Speech* (2013). We may recall his "burn[ing]" "tongue" in the passage from "Why Has Critique Lost its Steam" quoted above. More to the point here, it is what Latour attempts, under the label "diplomacy," in regard to, among other things, constructivist accounts of science in the Anthropocene (Latour 2015, "Diplomacy in the Face of Gaia").

II

I turn now to a second set of perplexities related to climate change: not efforts on the part of assorted devils to discredit evidence of its reality but the denial or disregard of that evidence on the part of large segments of the public. Current writings by environmentalists invoke a suite of psychological tendencies to explain these reactions—or, as one witty title has it, *Don't Even Think About It: Why Our Brains are Wired to Ignore Climate Change* (Marshall 2014).[94] A different angle on them is suggested by an important line of empirically informed inquiry in cognitive theory and philosophy of mind. For those pursuing this line of thought, the question is not, as in classical epistemology, how we, or the philosophers or scientists among us, arrive at the truth. The question is, rather, how humans generally come to believe what they believe or know what they think they know.

In traditional accounts, beliefs are commonly conceived as discrete, correct-or-incorrect propositions about the world ("the earth is flat," "the earth is round," and so forth) located in our minds. Beliefs might be better conceived, however, as systems of linked assumptions, ideas, images, and recollections and related perceptual and behavioral dispositions. This alternative view is suggested by several intellectual traditions. One is research and theory in empirical psychology, cognitive science, and neuroscience. A second, drawing on the first and on work in theoretical biology, is an influential approach in philosophy of mind.[95] The relevant accounts in each of these fields stress the *dynamic, interactive,* and *embodied* character of the processes and products of cognition: that is, of learning, remembering, figuring out, and planning, and of knowledge, memories, ideas, and beliefs. Increasingly in these fields, human cognition is identified not with so-called higher, rational, or computational processes occurring inside our heads but, rather, with our ongoing responsive interactions with our environments as fully embodied creatures.

Throughout our lives, we interact with our environments in ways that continuously modify our bodily structures and how they operate; and these modifications affect our subsequent interactions with our environments, both what we perceive and how we act. It is not only that our bodily structures and their operations define what we can "detect" about the world, but also that the world that each of us can *act on* and *be acted upon by* is a specific perceptual and behavioral niche. Through the very process of living, moving about in, and exploring the world, we, like other creatures, continuously form cognitive constructs of the features of those niches; and, in turn, we, like other creatures, are continuously shaped—bodily and thus also cognitively—by the features of the niches with which we interact. Given the reciprocal dynamics just described, we can never become pure spectators of the world, observing a reality altogether independent of us. This is not to say, however, that we are *locked out* of "the real world." It is to say, rather, that, like other creatures, we are inextricably *interlocked with* the real-as-can-be world. The objective reality posited in realist and neorealist ontologies is commonly evoked as "out there," as a realm of Being distinct and apart from the humans who, as it were, just happen to inhabit it. Our operative realities, however, that is, the perceptual and behavioral niches with which we can and do interact and which can be consequential for us, are not distinct from the humans—philosophers and scientists included—who inhabit them. On the contrary, they are continuously realized, brought forth and made real for us, through the very processes of our living in them.

Much of what we experience as external to ourselves and speak of as "reality" can be understood as the relatively stable, provisionally workable configuration of objects, forces, and processes that we continuously construct through our ongoing, effective-enough interactions with our experiential environments. The strong sense we may have of the simple out-there-ness of "what's out there" can be seen, accordingly, not as an undeniable intuition of an ontological given but as the thoroughly contingent product of complex processes of perceptual and behavioral coordination. These processes tend to issue in relatively similar patterns of response to relatively recurrent and stable conditions in our experience, and they are thus relatively predictable and reliable in their effects. We may appreciate, then, the experiential sources and apparent experiential confirmation of classic realist ideas. But we need not endorse those ideas in their classic versions. Specifically, we need not presume any isomorphism or other form of matching or mirroring between what we experience as and name "reality" and the putative objective properties of a presumptively independent universe. And we need not presume that the effectiveness of our individual actions or communally developed technologies depends on our having correct representations of a universe altogether apart from us in our minds, our science textbooks, or our engineering manuals.

The view of human cognition outlined above was developed relatively independently of the tradition of constructivist epistemology associated with the history and sociology of science. Both, however, reject classic rationalist accounts of knowledge; and the alternative accounts of knowledge and belief developed by each are mutually supportive. As noted in previous chapters, one of the earliest articulations of constructivist epistemology is Ludwik Fleck's *Genesis and Development of a Scientific Fact* (1979 [1935]). A scholar of medical history and shrewd amateur sociologist as well as a practicing biologist, Fleck sought to describe the formation of scientific knowledge in relation to the more general dynamics of human cognition at both the individual and the sociohistorical levels. Of particular interest here is his analysis of what he calls "belief systems" and what he describes as their "tendency to inertia." He writes: "Once a structurally complete and closed system of beliefs consisting of many details and relations has been formed, it offers tenacious resistance to anything that contradicts it" (27, translation modified). Fleck relates this tendency to what he describes as the ongoing "mutual attunement" of beliefs, perceptions, and material practices among the members of an epistemic community or, in his term, "thought collective." Through this process, collectives develop characteristic "thought styles," which is to say, shared habits of perception, interpretation, and explanation.

Several features of belief systems, so understood, are relevant to climate skepticism and denialism. One is the dense interdependence of their elements. For example, different sets of communally shared assumptions may dispose us to see—perceive and interpret—a severe drought as evidence of global warming, or as part of a normal weather cycle, or as punishment by the gods. And, reciprocally, our interpreting a drought in one or another of these ways will reinforce those assumptions and, with them, our continued disposition to perceive and respond to comparable events in just those ways. This ongoing positive feedback loop gives considerable stability to our belief systems. By the same token, however, it makes those systems strongly resistant to change.

A second relevant feature of belief systems, so understood, is their social constitution and institutional maintenance: that is, the fact that they are both formed and stabilized through our ongoing interactions with, among other things, other people, especially members of our particular epistemic communities. A good part of what re-stabilizes a conviction (for example, a belief that the earth is flat) in the face of potentially destabilizing experiences (for example, some clever fellow's demonstration that the earth is round) is the renewal of the collective social practices through which that conviction was formed in the first place—practices, it must be stressed, that are not only verbal but embodied and also, usually, materially consequential as well. If a belief system remains stable among the members of some collective, it is because the perceptual and behavioral dispositions associated with it allow its members to operate effectively in whatever domains of thought and practice are significant in their lives. Thus, convictions or beliefs that are clearly mistaken or irrational in relation to currently prevailing and sophisticated understandings of the world—for example, the conviction that droughts are divine punishments—may nevertheless serve, or have served, the members of a particular community well enough in the only worlds they actually inhabit/ed.

This perspective on belief systems allows us to appreciate some aspects of so-called belief-perseverance not always recognized by those who invoke the tendency to explain stubborn skepticisms regarding climate change. First, the tendency in question (I call it "cognitive conservatism" [Smith 1997, 50-51, 84-86]) is not a flaw in our mental makeup or a maladaptive hangover from our evolutionary history. Rather, it is a powerful but ambivalently operating—sometimes good, sometimes bad—feature of human cognition. Cognitive plasticity is obviously necessary for any creature that survives, as humans do, by its ability to learn and to modify its behavior responsively. But so also is the *countertendency*, that is, our ability and inclination to retain our beliefs beyond the occasion of their formation. Much of intellectual life and intellectual history can be understood as the interplay of these complementary tendencies and their assessment from different perspectives. Thus, given our well-developed capacity to learn from others, we can be duly "informed" and "enlightened" but also, as we say (usually of other people), "duped" or "indoctrinated." And, given our ability and inclination to hold fast to what we believe, we can be stubborn in our attachment to error or, as we say (usually of ourselves), steadfast in our defense of truth.

I would also stress, accordingly, the *generality* of cognitive conservatism. The tendencies associated with it—rationalization, dissonance-avoidance, confirmation bias, and so forth—are not restricted to the naive or the uneducated. Physicians hold fast to their diagnoses, scholars to their glosses. Philosophers discredit the intellectual competence of challengers; scientists discount anomalies as flukes. These tendencies are routinely attributed to climate skeptics and climate denialists. We should not be surprised if they were found among environmentalists as well.

III

This last possibility recalls Latour's suggestion, in the passage quoted earlier, that those engaged in critique should examine their own attitudes and practices. "Or should we rather bring the sword of criticism to criticism itself and do a bit of soul searching?," he asks, and adds: "Nothing guarantees, after all, that we should be right all the time." The suggestion has been taken as urging a wholesale abandonment of practices of critique and has spawned, accordingly, a good deal of rather indiscriminate anti-critical declaiming to that effect.[96] Latour has his own quarrels with some types of critique and often writes—indiscriminately enough—of "the critical spirit" or "the critical mind." But the soul-searching he recommends here can be understood differently: not as critical suicide but, rather, as a type of principled reflexivity urged and often pursued in science studies. Latour himself has urged and pursued it under the term "symmetrical anthropology." Thus he admonishes field anthropologists to "come home" and examine Western, modern beliefs the way they examine the beliefs of other cultures. And, as discussed in chapter 4 above, he illustrates the practice by describing the facts constructed and accorded authority in Western science in pointed parallel with the fetishes constructed and worshiped in the past by Gold Coast natives (Latour 2010, "On the Cult of the Factish Gods").

A recent "ontological turn" in anthropology is of related interest. The development is usually described as a determination by ethnographers and theoretical anthropologists both to "take seriously" the cosmologies of indigenous peoples and to question seriously their own ontological assumptions.[97] There may be no direct historical connections, but one can see this self-disciplining turn as a radicalization of the more general commitment to explanatory symmetry and reflexivity put forth in the sociology of science in the 1970s and sustained more generally in constructivist studies of science and technology (STS). As discussed in previous chapters, the aim of those twin commitments was to avoid the commonly lopsided—self-flattering and strongly presentist—explanations of scientific achievement familiar in rationalist epistemology and triumphalist intellectual history. And, as I have argued throughout, although the practices associated with those commitments are certainly relativistic in some respects, they are neither foolish nor objectionable but, on the contrary, can be especially valuable in dealing with conceptual and ethical perplexities, including the kind that concern us here.

An example relevant to climate change is a study by Arlie Russell Hochschild titled *Strangers in their Own Land* (2016). A social psychologist, Hochschild sought to understand why people living in some of the most petro-polluted regions of the United States vote regularly for candidates who support both the oil-extracting, chemical-producing industries responsible for the pollution and the free-market practices that have destroyed their natural environment and threaten their homes and health. She calls it "The Great Paradox." Focusing on a coastal region in Louisiana, Hochschild interviewed a good number of such people, following them in their daily routines and speaking with them in their homes, churches, and workplaces. She describes the study as involving efforts to surmount "the wall of empathy" that separated her, a self-described liberal and progressive, from the men and women—all professed members of the conservative Tea Party—whose lives and views she documented: factory workers, fishermen, gospel singers, small-business owners, local officials, and so forth. Hochschild's determination to report their views impartially is evident throughout the book. Of particular interest here are her efforts to find counterparts, in her own sentiments, to those expressed by her subjects and also to find parallels to them in feelings, views, and values commonly expressed by her liberal friends and associates. I return below to some of the Tea Party members' views and sentiments. What I want to note here are both the symmetry and the reflexivity displayed in those efforts: forms of intellectual and ethical good practice that can bring us, as they evidently brought Hochschild herself, to insights not otherwise readily available and to understandings of ourselves as well as others that can be politically as well as psychically valuable.[98]

Recent thought about these issues is inevitably shadowed by the coincident political dramas unfolding in the United States. The realities evoked in Hochschild's book are especially haunting. Many oppositional slogans, "Love trumps hate," "We have nothing to fear but fear," and so forth, are probably beside the point. The most politically significant views that Hochschild encountered among her Tea Party subjects were born not of fear or raw hatred but, rather, of nostalgia and deep resentment. While she does not use the latter term, her reports, like those of many other observers, suggest that, along with the sense of a better world lost, a sense of injured merit—and, in effect, of injustice—may be among the most powerful motivators of current rural and working-class politics in the United States.[99] The values and attitudes that Hochschild describes among the people she studied are largely the homely ones familiar to observers and social historians of the American South and Midwest: personal independence, in-group loyalty, hard work, and pride in status. Louisiana Tea Party voters are not resentful of their petro-industry employers, to whom they are grateful for jobs and for whatever measure of local prosperity they see. Nor are they resentful of superrich businesspeople, whom they admire for their industriousness and success. They are resentful of people and groups whom they see as getting special attention or getting ahead unfairly—largely blacks, gays, and immigrants—and perhaps most deeply of those, "liberals" and "elites," whom they see as putting those people and groups ahead while condemning *them* as backward, bigoted, or worse. The Great Paradox is hardly paradoxical at all.

Some reflections suggest themselves accordingly. It is probably not the "postmodern" critique of rationalist-realist epistemology that makes rural, working-class voters in the United States ignore climate science. It may not even require industry-funded campaigns to win and sustain their support for anti-regulation candidates. It may be enough that they see environmentalism, not altogether incorrectly, as a liberal-elite agenda and that global warming remains, for them, a distant and abstract concept.[100] At least two important lessons seem to follow. One is that hopes for a grassroots movement of resistance to the extraction of fossil fuels must be checked against the force of such views and of the political realities to which they give rise. Here as elsewhere, malice can trump prudence and resentment can trump everything. The other is that, to be effective, environmentalist arguments, like arguments against the devil quoting scripture, must be attuned to the conceptual idioms of their actual audiences and calibrated to their operative realities.

IV

In describing, above, the generality of belief-persistence, I noted that the tendencies associated with it might be found among environmentalists as well as among climate skeptics and denialists. What I had in mind was the persistent hope or conviction, common among the writers cited here, that widespread recognition of the perils involved in global warming will lead to the emergence of a climate-conscious and climate-conscientious human collective, either a major new coalition or, in effect, an ingroup of humanity as a whole.

Those voicing such hopes often invoke past cultural shifts and large-scale movements as precedents. Thus appeals are made to abolitionism, feminism, the anti-Vietnam War movement, and the acceptance of gay marriage, with the French or Russian revolutions and perhaps, for some, the rise of Christianity as shadow models of historical success.[101] But such appeals must ignore or downplay the significantly different forces and circumstances that exist here: notably, the diffuseness, complexity, and relative impersonality of the harms involved, the magnitude of the immediate costs or sacrifices required, and the uncertainty and abstractness of the benefits, if any, to be gained. In a recent book, *The Great Derangement*, Indian writer Amitav Ghosh, noting a range of discouraging historical precedents and comparably discouraging current political and economic conditions, writes (with a perhaps unwitting double emphasis): "The refusal to acknowledge these realities sometimes lends an air of unreality to discussions of climate change" (Ghosh 2016, 145).[102]

Other differences are equally important. The varied phenomena that we sum up as climate change have aspects that make due recognition both difficult to arrive at and difficult to communicate. Some of the relevant phenomena are violent and manifest but local and transient; others are large and widespread but slow and not readily perceptible; all are exceedingly unevenly distributed globally and even regionally. The relevant information is technical, complex, and not easily absorbed or retained by individuals, even those with specialized training. Such information is also neither ready to hand or easy to come by, and much of the data is, in fact, disputed among experts. Those in the past who theorized or summoned large-scale uprisings did not confront and probably could not have imagined the combined aggregating, stupefying, and disaggregating effects of the hyper-mass media, which, even as they enable the mobilization of multitudes, sharpen and polarize group differences and isolate communities from potentially illuminating interactions with others.

As emphasized by environmentalists, there is a virtual consensus of informed scientists regarding the destructive, destabilizing effects of anthropogenic climate change, and one could claim a virtual consensus of academics, intellectuals, and thoughtful, well-intentioned people regarding the need for a set of appropriate responses. This statement of double consensus, however, obscures a number of sticky issues. What responses *are* appropriate? By what methods, in what assemblies, and with what authority, will that appropriateness be determined, and in accord with which—or whose—interests, views, and priorities? Latour's essay, "Why has Critique Run out of Steam?," is often invoked, here as elsewhere, as urging us (presumably thoughtful, well-meaning academics and intellectuals) to turn from "matters of fact," meaning, perhaps (among other things) the detailed data of climate science, to "matters of concern," meaning, perhaps (among other things), the actual quotidian threats involved. Given, however, significant differences of situation as well as multiple, divergent interests among "us" (certainly among humans generally and even just among thoughtful, well-meaning academics and intellectuals), it is not clear, especially when equally justifiable interests conflict, which matters of concern should concern us most. *Humanity?* Every one of us? If not, then who? *The biosphere?* Every nook, niche, and creature? If not, then which ones? *Future generations?* How far into the future? If not thousands of years, then how many?

It is also not clear what form responses can take that would be on a scale great enough to have the required or desired effects. Fossil fuels should, of course, be replaced by renewable sources of energy. But where, when, by what technical, political, and/or economic means, and through what social agencies? Are geo-engineered protections or mitigations (cement pilings, floodgates, cloud seeding, and so forth) altogether to be disdained, as suggested by those intent on purely socio-economic and/or spiritual transformations? If so, then when exactly should we give up driving cars and having children, and through what form of incentive and/or compulsion, administered and monitored by what governmental agencies? Duly detailed attention to such questions might better prepare us for the conflicts and losses thereby indicated and perhaps encourage the compromises and sacrifices required by all for the survival of anything.[103] Conversely, without acknowledgment of these and other comparably difficult issues, environmentalists' summons to responsible action becomes only too readily ignorable.

Many scientists and environmental activists write as if geophysical facts will triumph of themselves or will do so when the machinations of "merchants of doubt" are exposed. Often overlooked is the significance of the variable conditions of the reception of those facts, and also the significance, for different listeners, of how the accuracy of that information is authorized and how such exposures are mediated. Recently and especially as spurred by climate change, historians and political scientists have joined sociologists of science in observing the limits of the so-called "deficit model" of science communication, in which scientists are seen primarily as inherently credible deliverers of accurate information to an ignorant public (Howe 2014, Simis et al 2016, Jasanoff and Simmet 2017). Environmental activists sometimes suggest that the Internet or social media amount to vastly expanded, presumably democratic, public spheres. But, as noted by most media scholars and commentators, these electronic venues have tended to become more or less tightly sealed social enclosures, as, of course, more familiar concrete spaces (pubs, squares, and town halls included) could also be. The "public sphere" of Critical Theory was always an idealized concept.

Writers voicing hopes for the emergence of an environment-conscious global collective also appeal to allegedly universal human values or natural impulses as dependable motivators. Thus British environmental activist George Marshall writes of the universality of parental care for children and other "nonnegotiable sacred values" (Marshall 2014), and American environmental historian Jedidiah Purdy describes the "self-restraint" that follows from a recognition of personal "guilt" and thus "responsibility" (Purdy 2016).[104] Others write of a "treasuring" of the natural world and life-forms awakened by encounters with natural beauty and sharpened by art or the due study of biology (Wilson and Kellert 1993). The universality and naturalness of such impulses may be doubted, however, along with the political effectiveness of such recognitions and treasuring as may, indeed, occur among various, but probably quite limited, groups of people. Those invoking such allegedly endemic responses and moral norms, parental love included, must disregard their well-attested cultural variability and also their evident failure to prevent persistent patterns of rather nasty behavior among humans everywhere. I remarked above that climate change doesn't change *everything*. It doesn't change *everyone*, either.

The difficulties involved in such appeals and arguments are illustrated in a recent book, *Defiant Earth* (2017), by Australian environmental ethicist Clive Hamilton. Hamilton calls for "a new anthropocentrism" anchored in a recognition of our unmatched achievements ("marvels of intellect and culture") joined with our equally unrivalled power to destroy. He goes on to argue that because, unlike other creatures, humans have the freedom to choose, we have the duty to repair or avert the environmental damages and threats caused by the unbridled exercise of our powers. Hamilton describes these ideas as "the defining truths of the age" (55), but his highly abstract moral logic is not likely to transform the behavior of many people. At the same time, the theologically tinged image of humans as demigods obscures an arguably more useful recognition of a range of relevant human attributes, including some reliable, though strictly sublunary, creative capacities.

In addition to the forms of denialism noted above, there is an inclination among environmentalist writers to suggest that the failings of *dominant* views and values (notably Western, modern, materialistic, and male-associated) are *exposed* by global warming and will be duly humbled while the benefits of *marginal* views and values (notably non-Western, traditional, spiritual, and female-associated) are *redeemed* by it and will be duly elevated.[105] There is sometimes a grim satisfaction to these suggestions of reversals. One could call it Anthropocene *Schadenfreude*: all may go down in darkness, but the despised and rejected—those who suffered under capitalism and imperialism, who railed against secularism and modernity, and who predicted catastrophe and ruin—will, as one writer puts it, have "the last laugh."[106]

Futurology is a high risk sport for amateurs and must be, for professionals, an increasingly thankless task. It is exceedingly hard to predict how the complex and highly differentiated effects of climate change will play out in the long run or even in the middle run. As I have suggested, there is good reason to doubt that those effects will be countered by the united efforts of a duly enlightened global human community. What seems more likely is the ad hoc emergence of local coalitions in the face of more or less urgent local threats. Barbarism may not be the only alternative to socialism, as Rosa Luxemburg believed and as Isabelle Stengers (2015) echoes. But it also seems likely that, as critical resources (habitable shelter, cultivatable land, food, water, energy, and so forth) become increasingly scarce, there will be struggles, more or less barbarous, between different groups.

My aim here has certainly not been to discredit the efforts of environmental advocates and activists. I admire greatly the scholars and writers whose works I have been citing and would want to share their most optimistic views. My aim, rather, has been to articulate some of the problems I see associated with these efforts and to suggest some alternative ways to respond to the concerns we share. It comes down, perhaps, to advising more relevantly informed imaginings and goals and to suggesting that we grant not, of course, the equal validity of all views or the equal consequentiality of all realities but the equal *force* of all views and realities for those who hold and inhabit them and, therefore, their substantial claim on the attention of those of us who—on whatever terrain, intellectual or practical, ethical or political—must operate in relation to them.

It is just as well that climate change does not change everything or everyone. Adaptive as well as maladaptive capacities will no doubt continue be displayed. At least some past wisdom is likely to remain a resource in the face of large-scale historical obliviousness. Even as realities change radically, humans will probably continue to form collectives, continue to instruct and to learn from one another, and, we may hope, operate in many places more or less effectively—and, sometimes, less rather than more barbarously.

Notes

Notes to Chapter 1

1. To address a contemporary concern: there is nothing anthropocentric in this view. In the accounts referenced here, the same is understood to be true of the relation between the processes and products of cognition in all organisms: that is, specification, articulation, and realization through embodied dynamic interaction.

2. See, e.g., Fleck 1935, Kuhn 1962, Feyerabend 1975 and 1985, Gibson 1979, Maturana and Varela 1980, Knorr Cetina 1981, Latour 1988, Pickering 1995, Noë 2004, Thompson 2010, Hutto and Myin 2017. Constructivist challenges to classical epistemology are by no means recent in origin. Some have been part of the philosophical tradition since Protagoras; others can be traced without difficulty to the ideas of Nietzsche, Wittgenstein, Heidegger, William James, and John Dewey. For some of the lineages, see Golinsky 1995. For a sample of the range of contemporary constructivist approaches and a set of efforts at a comprehensive definition, see Becerra and Castorina 2018.

3. See, e.g., Boghossian 2006. Boghossian casts constructivism as a politically motivated "fear of knowledge" on the part of members of socially disadvantaged groups and the philosophically disadvantaged academics who speak for them.

4. For discussion of what I call the Egalitarian Fallacy and other improper charges and inferences, including quietism, nihilism, anything-goes-ism, and self-refutation, see Smith 1988, 150-184; 1997, 1-36, 73-87; 2005/6, 18-45. For equivocal invocations and/or disavowals of relativism, see Smith 2005/6, 85-107.

5. For examples of such practices in the "science wars," see Gross and Levitt 1998, Sokal and Bricmont 1998, and Koertge 1998, 3–6, discussed in Smith 2005/6, 115-127. For the display of such practices in a purported exposure of the "unpalatable relativism" of constructivism, see Boghossian 2002, discussed in Smith 2002 ("Reply to an Analytic Philosopher").

6. See, e.g., Harré and Krausz 1996, 75, 100, and 112-13.

7. For related discussion focused on invocations and denials of the Holocaust, see Smith 1997, 23-36. For comparable arguments regarding invocations and

denials of climate change and comparable questioning of the assumptions involved, see the exchanges between Baker and Oreskes 2017 and Fuller 2017.

8. For important early accounts of the developments, see Pickering 1992.

9. For an informed attempt to sort out the affinities and differences, see Jensen 2017.

10. For the former, see, e.g., Boyer 2001. For the latter, see, e.g., Haught 2003.

11. "[W]e may have to abandon, for our part, the very notion of 'Belief'" (Latour 2013, 14). In an entry on the term "Belief" in the online version of the book (AIME), Latour writes: "The inquiry proceeds from an original form of agnosticism which consists not of disbelieving but of leaving entirely to one side the notion of belief in order to treat veridiction conflicts between modes" (AIME, accessed 23 December 2017).

12. Latour, AIME, entry on "Belief." Accessed 23 December 2017.

13. See, e.g., the glib dismissals of religion by a number of philosophers and scientists at the symposia *Beyond Belief* 2006 and 2007.

14. For an early pragmatist version, see Dewey 1958. "To see the organism in nature," Dewey wrote, "is the answer to the problems which haunt philosophy. And when thus seen [the organism] ... will be seen to be in [nature], not as marbles are in a box but as events are in history, in a moving, growing never finished process" (295). For a survey of such views of cognition and a discussion of the variants, see Hutto and Myin 2017, 1-53.

15. To say that cognitive processes and products are not reducible to two distinct types is not to say that they are homogeneous. On the contrary, what is important here are both the general heterogeneity of what we call our "beliefs" and the individual distinctiveness of our mental lives. For related discussion, see Smith 2010, 69-80.

16. For the continuity of the dispositions and activities commonly associated with religion and those involved in art and politics, see Burkert 1998. On the evident continuity of anthropologists' understandings of "religion" and "culture," see Smith 2009, 89-94.

17. For a comparable characterization, see the text version, Latour 2013, 184-185.

18. As observed by most scholars of the subject (though characteristically disputed by monotheologians), the term "religion" requires comparable pluralizing in comparable controversies.

Notes to Chapter 2

19. See Harré and Krausz 1996 for a taxonomy of relativisms and an attempt, as they write, "to extricate and examine the arguments, abstracted from

sources ancient and modern, that have been offered for and against the main varieties" (1). See Baghramian 2004 for the idea of a "family" of claims, positions, or doctrines. For an unhappy evocation of the supposed (and perhaps real enough) moral relativism of contemporary college students, see Thomson 2001, discussed in Smith 2001 ("Comment"). For a "postmodern relativism" identified by a failure on the part of various scholars to offer certain desirable emphases in their work, see Mohanty 1997 and the discussion in Smith 2005/6, 34-38.

20. On the early figures and what I discuss as "pre-post-modern relativism," see Smith 2005/6, 18-45.

21. These intentions are clear in the following passages by anthropologist Ruth Benedict and sociologist David Bloor:

> Any scientific study requires that there be no preferential weighting of one or another of the items in the series it selects for its consideration. In all the less controversial fields like the study of cacti or termites or the nature of nebulae, the necessary method of study is to group the relevant material and to take note of all possible variant forms and conditions .... It is only in the study of man himself that the major social sciences have substituted the study of one local variation, that of Western civilization. (Benedict [1934] 1959, 3)
>
> If [the sociologist's] theories are to satisfy the requirement of maximum generality they will have to apply to both true and false beliefs .... The sociologist seeks theories which explain the beliefs which are in fact found, regardless of how the investigator evaluates them .... The approaches that have just been sketched suggest that the sociology of scientific knowledge should adhere to the following four tenets .... (Bloor [1976] 1991, 5, 7).

22. I discuss the self-affirming circularity involved in such arguments in Smith 1997, 73-87 and 118-124.

23. Despite the quantities of logical ink poured over their heads, none of the major thinkers thus putatively refuted—for example, Feyerabend, Foucault, Derrida, Rorty, or Latour—has seen reason to acknowledge the alleged error of their thinking on the crucial issues.

24. I refer here to such principles as historical and cultural contextualization and explanatory impartiality and reflexivity, not to such (alleged) views as the utter uniqueness of every culture or the impossibility of cross-cultural generalizations.

25. For work on linguistic relativity, see Gumperz and Levinson 1996. For the historicity of concepts such as *science, knowledge,* and *objectivity,* see Giere 2006, Daston and Galison 2007, Shapin 2008. For a rigorous articulation of epistemological relativism, defended handily against standard philosophical

refutation-arguments, see Kusch 2002. For philosophical defenses of relativism as such and a useful overview of current philosophical treatments of the issues, see García-Carpintero and Kölbel 2008.

26. For examples of such denunciations, see, respectively, Wilson 1998, Boghossian 2006, and Ratzinger 2005. For an exegesis of "the dictatorship of relativism" in the Vatican example, see Smith 2007.

27. Atran's examples are singularly ill-chosen. Many lay readers as well as anthropologists would find such beliefs recognizable and have no trouble restating them in familiar, credible-enough terms. But his claim here is already strained. The cultural relativism of anthropologists does not typically arise from finding the beliefs of the people they study nonsensical as distinct from deeply different from beliefs generally accepted in Western cultures (see, e.g., Vivieros de Castro 2015).

28. See the discussion of dynamic, embodied, ecological, and enactive views of human cognition in chapters 1, 4 and 7. For further discussion of Atran's examples and argument, see Smith 2009, 10-19.

29. For extensively informed discussion of these general points, see Lloyd 2007.

30. Behaviorists and Jesuits famously spoke of the power of conditioning or early training to shape the minds of the young.

31. Prominent among the alternative approaches are developmental systems theory (see Oyama et al 2001) and, as noted above, accounts of cognitive processes that stress their dynamic, interactive features (e.g., Núñez and Freeman 1999, Noë 2004, Melser 2004).

32. Fuller himself pointedly rejects the idea that the pursuit of sociology of science entails a refusal to offer judgments on public issues involving science. For an exchange between Fuller and fellow sociologists of science regarding his testimony in Kitzmiller v. Dover, see Lynch et al 2006.

33. In a subsequent commentary on his participation in the Dover trial, Fuller recognizes the political value of a contingency-conscious relativism. Noting that science studies "have a weak public presence and a history of being treated as pawns by more powerful players," he writes: "Had I been more of a relativist, presumably I would have taken heed of these features of the situation and refrained from offering my services" (Fuller 2009, 116).

34. Such a disavowal occurs, for example, at the end of an analysis of various contemporary proposals for responding to the use or abuse of drugs by Jacques Derrida. The analysis itself is thoroughly mindful of contingency and explicitly committed to symmetry and evenhandedness. Derrida writes:

> Depending on the circumstance ... the discourse of "interdiction" can be justified just as well or just as badly as the liberal discourse .... Since it is impossible to justify absolutely either the one or the other of these

practices, one can never absolutely condemn them. In an emergency, this can only lead to equivocations, negotiations, and unstable compromises. And in any given, progressively evolving situation, these will need to be guided by a concern for the singularity of each individual experience and by a socio-political analysis that is at once as broadly and as finely tuned as possible. *I say this not to avoid the question any more than I do to argue for relativism.* (Derrida 1995, 239, emphasis added)

Martin Hägglund quotes the above passage following his sympathetic explication of a form of political thinking that he terms "hyperpoliticization" (2008, 234). Doubling the both the arguably relativistic observations and the explicit refusal of the presumed intellectual weakness, Hägglund writes:

> For a hyperpolitical thinking, nothing (no set of values, no principle, no demand or political struggle) can be posited as good in itself .... Rather than having an ultimate legitimacy, it [any political order] can be challenged on the basis of what it does not include and must remain open to contestation because of its temporal constitution. *To assert such a condition is not to give in to relativism"* (184–85, emphasis added).

These disavowals of an undefined relativism appended to rigorous articulations of strenuously contingency-affirming, absolutism-rejecting views raise the question of what, in each case, "relativism" and either "argu[ing] for" or "giv[ing] in to" it are understood to mean. More generally, one might ask: On the basis of what specific considerations of intellectual history, or as the result of what individual experiences of usages of the term, are such disavowals thought necessary?

Notes to Chapter 4

35. The original version of this essay was prepared for a symposium titled Recomposing the Humanities with Bruno Latour.

36. The account summarized here is initially developed in Latour and Woolgar 1986 and Latour 1987 and 1988. It is further elaborated and elucidated in Latour 1999 and 2005 (*Reassembling the Social*).

37. See chapters 5 and 6 in this volume.

38. Latour's usage of the term *Moderns*—and, in connection with it, either "we" or "they"—varies widely, and the specific reference of the term in his work tends to be elusive. Most generally and neutrally, it seems to mean something like educated, post-Enlightenment, more or less secularized Westerners. Throughout his writings, however, and in tones ranging from affectionate irony to bitter sarcasm, Latour depicts the members of this group (or, in the ethnographic conceit of Latour 2013, this "tribe") as fundamentally benighted, self-ignorant, and arrogant: mistaken about the constitution of their world, mistaken about their own motives and values, and given to airs

regarding those they regard as unenlightened. Since he gives few specific examples of individual Moderns, either historical or contemporary, readers will be inclined to understand the references in accord with their particular sense of Western history and their more general intellectual and cultural tastes and, of course, distastes.

39. See Latour and Wiebel 2002. The recurrent quote marks are Latour's.

40. See, for example, Latour's postface to the French translation of *Genesis and Development* (Latour 2005, "*Transmettre la syphilis, partager l'objectivité*") and Latour 2013, 91.

41. As noted in chapter 1 above, what Latour would banish is not the term "belief" as such (he acknowledges its innocuous usages) but its invidious or patronizing invocation, largely in relation to religious ideas. His efforts "to slip in between" what he calls "epistemological questions and ontological questions" reflect his related effort, in establishing the respect-worthiness of divinities, to escape the choice between a dubious claim of objective existence for such beings and an unwanted ascription of their existence to (mere) subjective belief.

42. Latour stresses that it is also a mistake, though of a different kind, to appeal to a putative correspondence-to-reality to explain the efficacies of the natural sciences.

43. See, for example, Tim Howles's rejoinder to Jan Golinski's appreciative but distanced reading of "'Thou Shall Not Freeze-Frame'" (Golinski 2010). Howles writes:

> Religious people, Golinski thinks, "will still want to insist on the ontological reality of the things they believe in and will not be happy to have their religion reduced to the manipulation of signs that lack any reference to the real world." ... However, in the light of this chapter [i.e., chap. 11 in *Modes of Existence*], I suggest we can put Golinksi's claim to bed as unfounded. Latour does not lead us into the realm of apophatic theology and the beings of [REL(IGION)] are not to be taken as merely Feuerbachian projections. There is ballast to Latour's theology (Howles 2014).

44. See, e.g., the account of these writings by theologian Adam S. Miller (2013). Miller's style is, like Latour's, highly allusive and, in his case, also exceedingly gnomic.

45. The scope of the concept and the meanings of the term "religion" are, of course, extensively contested. On the concept, see Saler 1993 and Dubuisson 2003; on the term, see J. Z. Smith 1998.

46. My term "religion-proper" here and below refers to the mode that Latour names "Religion" and identifies, usually indirectly but always recognizably, with Christianity.

47. Latour's references to non-Catholics or non-Christians tend to join them together in terms that are, at best, vague: for example, "those outside" versus "those inside" the Church or, sometimes, "the indifferent" versus "the faithful."

48. In an entry on "Empiricism" in AIME, Latour writes: "Rereading James allows us to take radical empiricism as a watchword, but the phrase 'radical empiricism' takes on a more developed sense in AIME... AIME's radicalism is even more extreme."

49. "There is a constant risk," Latour writes, "of interpolating, confusing the two, failing to respect the contrasts. To care for is not to save. To initiate the circulation of psychogenics is not at all the same thing as letting oneself be overwhelmed by angels" (Latour 2013, 304).

50. For a range of examples of such efforts, see Kurz 2003.

51. On the historical connections, see Brook 1991, Bowler and Morus 2005, 341–66, and Harrison 2015. On the continuing connections, see Noble 1992.

52. I discuss the limited success of theological attempts at reconciliation in Smith 2009, 95–120.

53. See, for example, the chapter in Latour 1999 titled "The Historicity of Things: Where Were Microbes before Pasteur?" (145–73) and, with notably different concerns and emphases, the article titled "Charles Péguy: Time, Space, and *le Monde Moderne*" (Latour 2015).

54. Elements of the tale recur throughout Latour's writings. For recent instances, see Latour 2015 ("Charles Péguy"), 50, and Latour 2017, especially the sixth lecture, 184-219.

55. Representations of modernity as lapse, loss, and degeneration are a staple of traditional religious moralism and social conservatism and recur in more sophisticated forms in current so-called postsecularist thought. See, for example, Macintyre 1981, Taylor 2007, Gregory 2012, and Pfau 2013.

56. See the entry on "Psychology" in AIME, from which the epigraph to this section is drawn.

57. For related discussion and examples of these approaches, see pp. 15-16 and 112-114 in this volume. I consider their implications for controversies in epistemology in Smith 1997, 37–51, 125–52. For their relevance to the understanding of beliefs, religious and other, see Smith 2009, 5–19.

58. The alternatives here are also commonly seen as the ("merely") "subjective" and the (putatively) "objective." As such, they are what Latour has sought,

in his writings on religion and more generally, "to slip in between" or, precisely, to finesse.

Notes to Chapter 5

59. Brooks continues: "In this matter the textbooks that had been put in their hands were almost useless. The authors had something to say about the poet's life and the circumstances of his composition of the poem under study .... But the typical commentary did not provide an induction to this poem—or into poetry generally. The dollop of impressionistic criticism with which the commentary usually concluded certainly did not supply the need" (593).

60. Although they rebelled against the dominance of biographical scholarship, they did not, as is sometimes said, "banish the author." As would be clear from any page of criticism by Empson, Eliot, or Ransom, authors were very much in evidence, but as artists, not as mere historical figures or biographical subjects.

61. See chapter 6, below, for examples and further discussion.

62. The echoes of Popper in Moretti's title ("Conjectures on World Literature") and in Jockers' concern with literary interpretations that cannot be refuted are not accidental. Popper's influence, direct or indirect, is evident throughout both their arguments.

63. For the longer story, see chapter 6 below and Smith 2005/6, 108–29.

64. Their quote marks on the latter term are not explained.

65. See Trumpener 2009. Moretti replies in the same issue of the journal.

66. In an instructive chapter in a volume of *The Cambridge History of Literary Criticism*, Wallace Martin notes that, by the 1950s, the New Critics themselves worried about the "proliferation of over-ingenious interpretations" and that some people felt that "New Criticism had dwindled into a pointless routine" (2000, 316). Martin adds: "Yet it is worth recalling that scholars had registered the same objection to scholarship in the 1920s; driven by the conventions of the disciplines to make a 'contribution to knowledge', writers [it was said] proposed 'preposterous interpretations.'" "Admittedly," he comments, "there had been a change in the provenance of interpretative activity. Instead of asserting that Bottom [in *A Midsummer Night's Dream*] was James VI, the modern exegete would discover an archetype or a paradox" (316).

67. "Cavalier" is the term Bush uses in a rather strained set of puns (1949, 17).

68. For an arresting follow-us-or-die set of views, see Saklofske et al 2012. The authors write:

Taking a wait-and-see attitude that cautiously preserves the status quo is akin to choosing an unnecessary slow death over the possibility of an innovative cure. In an era of budget crises, enrolment uncertainty, and an increasing lack of connections between university-level career preparation and professional practice, it would be foolish to ignore an opportunity to reinvent [the humanities] and reconsider existing paradigms and practices. Digital humanities represents an already-established movement away from the doom-inviting stasis of the secondhand conservatism of universities that know the Net Generation has come, and yet decline to build the education system Net Geners both want and need. (329)

Notes to Chapter 6

69. Wilson describes himself in the preface to *Consilience* as a former fervent Baptist turned fervent believer in science.

70. See, e.g., Moretti 2007, where Franco Moretti, promoting big data and digital methods, invokes Kuhn's *Structure* to explain how genres of the novel appear in discrete "cycles." As noted in chapter 5, however, his major reference for scientific method is Karl Popper, whose *Conjectures and Refutations: The Growth of Scientific Knowledge* (1968) informs the title and supplies many of the details ("hunches" and "hypotheses," "tests" and "experiments," "corroborations" and "falsifications") of Moretti's influential article, "Conjectures on World Literature" (2000).

71. See Smith 2005/6, 46-84 ("Netting Truth: Ludwik Fleck's Constructivist Genealogy").

72. See Liu 2013. Liu argues that the significance ("meaning") of digital humanities projects involves their ability—now, he suggests, quite limited—to satisfy our interest in what we generally call "meaning."

73. See Smith 2005/6, 130-52 ("Super Natural Science: The Claims of Evolutionary Psychology").

74. Not surprisingly, literary Darwinism has received a good bit of sharp critical attention from scholars in the humanities. See especially Kramnick 2011 and 2012.

75. For the former, see, e.g., Delbanco 2013. For the latter, see, e.g., the commissioned report on the humanities titled *The Heart of the Matter* (2013).

76. See e.g., Wilson 1998, 7, quoted above, and the dubious suggestion, by a promoter of such views, that "scholars in the literary humanities have struggled to achieve at least a semblance of the certitude possible in the sciences" (Fromm 2006).

77. Given a history of dichotomous distinctions between "true knowledge" on the one hand and "mere opinion," "mere belief," or "mere superstition" on the other, it is not surprising that controversies over the relations between the sciences and the humanities continue to be dominated by struggles over the term *knowledge*: who owns it, who can deliver it, whose kind is "genuine."

78. "Cultural" in these connections is generally understood as artistically or intellectually worthy as distinct from strictly physical, merely useful, merely commercially successful, or merely popular. Boundaries between such categories, however, are hard to keep clear, and classifications of individual genres and achievements—as in the cases of jazz, journalism, photography, gymnastics, cinema, or video-game design—are routinely subject to struggle and shifting negotiation.

79. Carroll, in an interview for the journal *Science* (Kean 2011, 655), remarks that "most [humanities scholarship] today" is "unable to contribute in any useful way to the serious world of adult knowledge." Philosopher Alex Rosenberg writes that good naturalists "cannot take [literary studies] seriously as knowledge" if scholars "transparently flout science's standards of objectivity." He adds, evidently enjoying being wicked: "That does not mean anyone should stop doing literary criticism any more than forgoing fiction. Naturalism treats both as fun, but neither as knowledge" (Rosenberg 2011). See note 77, above, for the term "knowledge."

80. Thus literary Darwinist Blakey Vermeule proposes, perhaps jokingly, that the neatly turned verse couplets of Alexander Pope can be explained the same way as the gaudy tail of the peacock: that is, as a display of a "handicapping" trait (here a time-consuming and otherwise useless talent for verbal wit) that evolved by sexual selection (Vermeule 2012).

81. See, e.g., the descriptions of "distant reading" and "macroanalysis" in chapter 5 above.

82. For a candid account of the occurrence of just such dissonances in a collaborative project, see Fitzgerald et al 2014.

83. For efforts by phenomenologists to bridge the first-person/third-person divide, see Petitot et al 2000.

84. For chastened responses, see Zunshine 2010. The contrast is to chest-thumping works like Boyd et al 2010.

Notes to Chapter 7

85. See http://media.mcgill.ca/en/content/climate-realism-international-colloquium.

86. In a subsequent work, Kitcher defends a "modest realism" that seeks to incorporate more innovative components (Kitcher 2002). His position on those components, however, remains equivocal.

87. For an engaging image of such an imagined glimpse, see the late 19th-century Flammarion woodcut (figure 1). Here a pilgrim evidently catches a view of the machinery behind the everyday world of trees and houses, day and night, sun, moon, and stars. Notably, aside from one hand, only his head ventures—or, perhaps, can gain admission—into that realm.

88. Such challenges to realism, at least when mounted by certifiable academic philosophers, are generally referred to in Anglo-analytic philosophy as "anti-realism."

89. For comparable ideas developed in science studies (STS), see Pickering 1995 and Gad et al 2015.

90. My reference is to the title of the important book, Klein 2014 (*This Changes Everything: Capitalism vs. The Climate*).

91. For the former, see, e.g., M. P. Lynch 2017 ("Kick This Rock: Climate Change and Our Common Reality"). For the latter, see, e.g., Manda 2003 and Lepore 2017.

92. See, e.g., Latour 2013 and 2017.

93. Most egregiously perhaps, it does not explain Latour's caustic allusions to deconstruction or to the training of students in the humanities. One may note, however, that the essay was originally a lecture given at the Stanford Humanities Center.

94. See also Hoffman 2015 and Stoknes 2015.

95. For important works exemplifying these approaches, see Gibson 1979, Varela et al 1991, Thelen and Smith 1994, Port and van Gelder 1995, Núñez and Freeman 1999, Noë 2004, Chemero 2009, Thompson 2010, Anderson 2014, and Hutto and Myin 2017. Barrett 2011 and Sharma 2015 offer original, engaging, and instructive introductions to them.

96. For incisive commentary, see Barnwell 2016.

97. For examples, see Charbonnier et al 2017. For informed commentary, see Jensen 2017.

98. Such insights and understandings are not, of course, thereby guaranteed. See M. Lynch 2000 for a cautionary analysis of the contingent operations of methodological reflexivity.

99. The sentiment described here is evidently powerful across the current political and demographic spectrum, and not only in the United States. We seem generally to see it as a sense of injustice when displayed by those we

ourselves see as victimized and as resentment when we are skeptical of the merit, injury, or idea of justice implicitly invoked.

100. Most of Hochschild's subjects see global warming as a remote problem. Some of them expect a different apocalypse, with a Rapture of the pious, in their own lifetimes.

101. See, for example, Klein 2014, Marshall 2014, Oreskes and Conway 2014, Stengers 2015, and Purdy 2016.

102. For comparable observations and extented analyses, see Clark 2015 and Lorimer 2017.

103. On the significance of varied perspectives on such issues and the lack of public forums for their discussion and negotiation, see Jasanoff and Simmet 2017.

104. Environment-conscious writer Amitav Ghosh suggests that a broad-based religious or quasi-religious movement might supply the needed transformative energy, but he does not appear to anticipate one (Ghosh 2016). Other climate activists are explicit in invoking theology, religion, or religiosity as a desirable or crucial element of a due awakening (see, e.g., Stephenson 2015). For a detailed account of invocations and covert appropriations of religiosity by environmentalists, see Nelson 2010.

105. The image of existing class hierarchies overturned when conditions of survival change radically is evidently a recurrent conceit. See, e.g., Pierre Marivaux's *The Island of Slaves* and J. M. Barrie's *The Admirable Crichton*, in both of which, after a shipwreck, practical-minded, able-bodied slaves or servants become the masters.

106. Such people, including writers of fantasy and science fiction—genres disdained by the literary establishment in favor of classically "realist" novels—are what Amitav Ghosh calls, at one point, "the losers." Alluding to several such works by non-Western writers ignored by Western critics but especially relevant to climate change, he comments: "But once again, the last laugh goes to that sly critic, the Anthropocene, which has muddied, and perhaps even reversed, our understanding of what it means to be 'advanced'" (Ghosh 2016, 80).

Works Cited

Anderson, Michael L. *After Phrenology: Neural Reuse and the Interactive Brain*. MIT P, 2014.

Atran, Scott. *In Gods We Trust: The Evolutionary Landscape of Religion*. Oxford UP, 2002.

Baghramian, Maria. *Relativism*. Routledge, 2004.

Baker, Eric, and Naomi Oreskes. "It's No Game: Post-Truth and the Obligations of Science Studies." *Social Epistemology Review and Reply Collective* 6, no. 8, 2017, pp. 1-10.

Baker, Eric, and Naomi Oreskes. "Science as a Game, Marketplace or Both: A Reply to Steve Fuller." *Social Epistemology Review and Reply Collective* 6, no. 9, 2017, pp. 65-69.

Barnwell, Ashley. "Entanglements of Evidence in the Turn Against Critique." *Cultural Studies,* vol. 30, no. 6, 2016, pp. 906-925.

Barrett, Louise. *Beyond the Brain: How Body and Environment Shape Animal and Human Minds*. Princeton UP, 2011.

Becerra, Gastón, and José Antonio Castorina. "Towards a Dialogue Among Constructivist Research Programs." *Constructivist Foundations*, vol. 13, no. 2, 2018, pp. 191-218, http://www.univie.ac.at/constructivism/journal/articles/13/2/191.becerra.pdf.

Benedict, Ruth. *Patterns of Culture*. Houghton Mifflin, 1959 [originally 1934].

Beyond Belief: Science, Religion, Reason and Survival. 2006, http://thesciencenetwork.org/programs/beyond-belief-science-religion-reason-and-survival.

Beyond Belief: Enlightenment 2.0. 2007, http://thesciencenetwork.org/programs/beyond-belief-enlightenment-2-0.

Bloor, David. *Knowledge and Social Imagery.* 2nd ed., U of Chicago P., 1991 [originally 1976].

Boghossian, Paul. "Constructivist and Relativist Conceptions of Knowledge in Contemporary (Anti-) Epistemology: A Reply to Barbara Herrnstein Smith." *The South Atlantic Quarterly*, vol. 101, no. 1, 2002, pp. 213-227.

———. *Fear of Knowledge: Against Relativism and Constructivism.* Oxford UP, 2006.

Boyer, Pascal. *Religion Explained: The Evolutionary Origins of Religious Thought.* Basic Books, 2001.

Bowler, Peter J., and Iwan Rhys Morus. *Making Modern Science: A Historical Survey.* U of Chicago P, 2005.

Boyd, Brian, Joseph Carroll, and Jonathan Gottschall, editors. *Evolution, Literature, and Film: A Reader.* Columbia UP, 2010.

Brook, John Hedley. *Science and Religion: Some Historical Perspectives.* Cambridge UP, 1991.

Brooks, Cleanth. "The New Criticism." *Sewanee Review*, no. 87, 1979, pp. 592–607.

———. *Community, Religion, and Literature: Essays.* U of Missouri P, 1995.

Brooks, Cleanth, and Robert Penn Warren. *Understanding Poetry.* Holt Rinehart and Winston, 1960.

Burdick, Anne, Johanna Drucker, Peter Lunenfeld, Todd Presner, and Jeffrey Schnapp. *Digital_Humanities.* MIT P, 2012.

Burkert, Walter. *Creation of the Sacred: Tracks of Biology in Early Religions.* Harvard UP, 1998.

Bush, Douglas. "The New Criticism: Some Old-Fashioned Queries." *PMLA*, 64, supplement, part 2, 1949, pp. 13–21.

Carroll, Joseph. "Three Scenarios for Literary Darwinism." *New Literary History*, vol. 41, no. 1, 2010, pp. 53–67.

Charbonnier, Pierre, Gildas Salmon, and Peter Skafish, editors. *Comparative Metaphysics: Ontology after Anthropology.* Rowman and Littlefield, 2017.

Chatterjee, Anjan. "Neuroaesthetics: A Coming of Age Story." *Journal of Cognitive Neuroscience*, vol. 23, no. 1, 2011, pp. 53-62.

Chemero, Anthony. *Radical Embodied Cognitive Science.* MIT P, 2009.

Clark, Timothy. *Ecocriticism on the Edge: The Anthropocene as a Threshold Concept.* Bloomsbury, 2015.

Cook, Richard, Geoffrey Bird, Caroline Catmur, Clare Press, and Cecilia Heyes. "Mirror Neurons: From Origin to Function." *Behavioral and Brain Sciences*, vol. 37, no. 2, 2014, pp. 177-92.

Davidson, Donald. *Inquiries into Truth and Interpretation.* Clarendon, 1984.

Davis, Garrick. "The Well-Wrought Textbook." *Humanities*, vol. 32, no. 4, 2011, pp. 22–25.

Delbanco, Andrew. *College: What It Was, Is, and Should Be*. Princeton UP, 2012.

Demeritt, David. "Science studies, climate change and the prospects for constructivist critique." *Economy and Society*, vol. 35, no. 3, 2006, pp. 453-479.

Dennett, Daniel. *Breaking the Spell: Religion as a Natural Phenomenon*. Viking, 2006.

Derrida, Jacques. "The Rhetoric of Drugs." *Points ... Interviews, 1974 – 1994*, edited by Elisabeth Weber, Stanford UP, 1995.

Dewey, John. *Experience and Nature*. Dover, 1958 [originally 1925].

Drucker, Johanna. "Humanistic Theory and Digital Scholarship." *Debates in the Digital Humanities*, edited by Mathew K. Gold, U of Minnesota P, 2012, pp. 85-95.

Dubuisson, Daniel. *The Western Construction of Religion: Myths, Knowledge, and Ideology*, translated by William Sayers, Johns Hopkins UP, 2003.

Eliot, T. S. *The Sacred Wood: Essays on Poetry and Criticism*. Methuen, 1920.

Empson, William. *Seven Types of Ambiguity*. New Directions, 1947.

Feyerabend, Paul. *Against Method: Outline of an Anarchistic Theory of Knowledge*. Humanities P, 1975.

———. *Philosophical Papers: Realism, Rationalism and Scientific Method*. 2 vols. Cambridge UP, 1985.

Fish, Stanley. "Must There Be a Bottom Line?" *The New York Times*, January 18, 2010, https://opinionator.blogs.nytimes.com/2010/01/18/must-there-be-a-bottom-line/.

Fitzgerald, Des, Melissa M. Littlefield, Kasper J. Knudsen, James Tonks, and Martin J. Dietz. "Ambivalence, Equivocation, and the Politics of Experimental Knowledge: A Transdisciplinary Neuroscience Encounter." *Social Studies of Science*, vol. 44, no. 5, 2014, pp. 701 – 21.

Fleck, Ludwik. *Genesis and Development of a Scientific Fact*. Edited by Thaddeus J. Trenn and Robert K. Merton, translated by Fred Bradley and Thaddeus J. Trenn, U of Chicago P, 1979 [originally *Entstehung und Entwicklund einer wissenschaftlichen Tatsache*, Basel, 1935].

Fromm, Harold. "Reading with Selection in Mind." *Science* 311, February 3, 2006, p. 612.

Fuller, Steve. "Letter to the Editor." *Isis*, vol. 100, no. 1, 2009, pp. 115-16.

———. "What are You Playing At? On the Use and Abuse of Games in STS." *Social Epistemology Review and Reply Collective* 6, no. 9, 2017, pp. 39-49.

———. "Veritism as Fake Philosophy: Reply to Baker and Oreskes." *Social Epistemology Review and Reply Collective* 6, no. 10, 2017, pp. 47-51.

Gad, Christopher, Casper Bruun Jensen, and Brit Ross Winthereik. "Practical Ontology: Worlds in STS and Anthropology." *NatureCulture* 3, 2015, pp. 67-86.

García-Carpintero, Manuel, and Max Kölbel. *Relative Truth*. Oxford UP, 2008.

Ghosh, Amitav. *The Great Derangement: Climate Change and the Unthinkable*. U of Chicago P, 2016.

Gibson, J. J. *The Ecological Approach to Visual Perception*. Houghton Mifflin, 1979.

Giere, Ronald N. *Scientific Perspectivism*. U of Chicago P, 2006.

Golinsky, Jan. *Making Natural Knowledge: Constructivism and the History of Science*. Cambridge UP, 1998.

———. "Science and Religion in Postmodern Perspective: The Case of Bruno Latour." *Science and Religion: New Historical Perspectives*, edited by Thomas Dixon, Geoffrey Cantor, and Stephen Pumfrey, Cambridge UP, 2010, pp. 50–68.

Gombrich, E. H. *Art and Illusion: A Study in the Psychology of Pictorial Representation*. Princeton UP, 1969.

Gould, Stephen J. *Rocks of Ages: Science and Religion in the Fullness of Life*. Random House, 1999.

Gregory, Brad S. *The Unintended Reformation: How A Religious Revolution Secularized Society*. Harvard UP, 2012.

Gross, Paul, and Norman Levitt. *The Higher Superstition: The Academic Left and its Quarrels with Science*. Johns Hopkins UP, 1994.

Gumperz John J., and Stephen C. Levinson, editors. *Rethinking Linguistic Relativity*. Cambridge UP, 1996.

Hägglund Martin. *Radical Atheism: Derrida and the Time of Life*. Stanford UP, 2008.

Hamilton, Clive. *Defiant Earth: The Fate of Humans in the Anthropocene*. Polity, 2017.

Harré, Rom, and Michael Krausz. *Varieties of Relativism*. Blackwell, 1996.

Harrison, Peter. *The Territories of Religion and Science*. U of Chicago P, 2015.

Haught, John. *Deeper than Darwin: The Prospect for Religion in an Age of Evolution*. Westview P, 2003.

Hayles, N. Katherine. "Cognition Everywhere: The Rise of the Cognitive Nonconscious and the Costs of Consciousness." *New Literary History*, vol. 45, no. 2, 2014, pp. 199-220.

The Heart of the Matter: The Humanities and Social Sciences for a Vibrant, Competitive, and Secure Nation. Report of the Commission on the Humanities and Social Sciences, American Academy of Arts and Sciences, 2013.

Hochschild, Arlie Russell. *Strangers in Their Own Land: Anger and Mourning on the American Right.* The New P, 2016.

Hoffman, Andrew J. *How the Culture Shapes the Climate Change Debate.* Stanford UP, 2015.

Howe, Joshua P. *Behind the Curve: Science and the Politics of Global Warming.* U of Washington P, 2104.

Howles, Tim. AIME Research Group site, http://aimegroup.wordp.com/2014/07/08/chapter-11-welcoming-the-beings-sensitive-to-theword/. Accessed 12 November 2016.

Hutchins, Edwin. *Cognition in the Wild.* MIT P, 1995.

Hutto, Daniel D., and Eric Myin. *Evolving Enactivism: Basic Minds Meet Content.* MIT P, 2017.

Jasanoff, Sheila, and Hilton R. Simmet. "No funeral bells: Public reason in a 'post-truth' age." *Social Studies of Science*, vol. 47, no. 5, 2017, pp. 751–770.

Jensen, Casper Bruun. "New Ontologies? Reflections on Some Recent 'Turns' in STS, Anthropology and Philosophy." *Social Anthropology*, vol. 25, no. 4, 2017, pp. 525-545.

Jockers, Matthew. *Macroanalysis: Digital Methods and Literary History.* U of Illinois P, 2013.

Kean, Sam. "Red in Tooth and Claw among the Literati." *Science,* 332, no. 6030, 2011, pp. 655-56.

Kenner, Hugh. "The Pedagogue as Critic." *The New Criticism and After*, edited by Thomas Daniel Young, U of Virginia P, 1976, pp. 36–55.

Kirsch, Adam. "Technology Is Taking Over English Departments: The False Promise of Digital Humanities." *New Republic*, May 2, 2014, https://newrepublic.com/article/117428/limits-digital-humanities-adam-kirsch.

Kitcher, Philip. *The Advancement of Science: Science without Legend, Objectivity without Illusions.* Oxford UP, 1995.

———. *Science, Truth and Democracy.* Oxford UP, 2002.

Klein, Naomi. *This Changes Everything: Capitalism vs. The Climate*. Simon & Schuster, 2014.

Knorr Cetina, Karin. *The Manufacture of Knowledge: An Essay on the Constructivist and Contextual Nature of Science*. Pergamon P, 1981.

Koertge, Noretta, editor. *A House Built on Sand: Exposing Postmodern Myths about Science*. Oxford UP, 1998.

Kramnick, Jonathan. "Against Literary Darwinism." *Critical Inquiry*, vol. 37, no. 2, 2011, pp. 315 – 47.

———. "Literary Studies and Science: A Reply to My Critics." *Critical Inquiry*, vol. 38, no. 2, 2012, pp. 431 – 60.

Kuhn, Thomas S. *The Structure of Scientific Revolutions*, first edition. U of Chicago P, 1962.

Kurz, Paul, editor. *Science and Religion: Are They Compatible?* Prometheus, 2003.

Kusch, Martin. *Psychologism: A Case Study in the Sociology of Philosophical Knowledge*. Routledge, 1995.

———. *Knowledge by Agreement: The Programme of Communitarian Epistemology*. Oxford UP, 2002.

Latour, Bruno. *Science in Action: How to Follow Scientists and Engineers through Society*. Harvard UP, 1987.

———. *The Pasteurization of France*. Translated by Alan Sheridan and John Law, Harvard UP, 1988.

———. *We Have Never Been Modern*. Translated by Catherine Porter, Harvard UP, 1993.

———. *Pandora's Hope: Essays on the Reality of Science Studies*. Harvard UP, 1999.

———. "Why Has Critique Run Out of Steam? From Matters of Fact to Matters of Concern." *Critical Inquiry*, vol. 30, no. 2, 2004, pp. 225-248.

———. *Reassembling the Social: An Introduction to Actor-Network-Theory*. Oxford UP, 2005.

———. "*Transmettre la syphilis, partager l'objectivité*." Postface to Ludwik Fleck, *Genèse et développement d'un fait scientifique*, translated by Nathalie Jas, Les Belles Lettres, 2005.

———. "'Thou Shall Not Freeze-Frame,' or, How Not to Misunderstand the Science and Religion Debate." *Science, Religion and the Human Experience*, edited by James D. Proctor, Oxford UP, 2005, pp. 27–48.

———. *The Making of Law: An Ethnography of the Conseil d'État.* Translated by Marina Brilman and Alain Pottage, Polity, 2010.

———. "Coming Out as a Philosopher." *Social Studies of Science,* vol. 40, no. 4, 2010, pp. 599–608.

———. "On the Cult of the Factish Gods." *On the Modern Cult of the Factish Gods,* translated by Catherine Porter and Heather MacLean, Duke UP, 2010, pp. 1–66.

———. *An Inquiry into Modes of Existence: An Anthropology of the Modern.* Translated by Catherine Porter, Harvard UP, 2013.

———. AIME. Online version of *An Inquiry into Modes of Existence,* http://www.modesofexistence.org/.

———. *Rejoicing: Or the Torments of Religious Speech.* Translated by Julie Rose, Polity, 2013.

———. "Biography of an Inquiry: On a Book about Modes of Existence." *Social Studies of Science,* vol. 43, no. 2, 2013, pp. 287–301.

———. "Agency at the Time of the Anthropocene." *New Literary History,* vol. 45, no. 1, 2014, pp. 1-18.

———. "Charles Péguy: Time, Space, and *le Monde Moderne.*" *New Literary History,* vol. 46, no. 1, 2015, pp. 41–62.

———. "Diplomacy in the Face of Gaia: Bruno Latour in Conversation with Heather Davis." *Art in the Anthropocene,* edited by Heather Davis and Etienne Turpin, Open Humanities P, 2015, pp. 43-58.

———. *Facing Gaia: Eight Lectures on the New Climatic Regime.* Polity, 2017.

Latour, Bruno, and Steven Woolgar. *Laboratory Life: The Construction of Scientific Facts.* 2nd edition, Princeton UP, 1986.

Latour, Bruno, and Peter Wiebel, editors. *Iconoclash: Beyond the Image Wars in Science, Religion, and Art.* MIT P, 2002.

Lepore, Jill. "Autumn of the Atom: How Arguments about Nuclear Weapons Shaped the Debate over Global Warming." *The New Yorker,* vol. 92, no. 47, January 30, 2017, p. 22.

Liu, Alan. "The Meaning of the Digital Humanities." *PMLA,* vol. 128, no. 2, 2013, pp. 409–23.

Lloyd, Geoffrey. *Cognitive Variations: Reflections on the Unity and Diversity of the Human Mind.* Oxford UP, 2007.

Lorimer, Jamie. "The Anthropo-Scene: A Guide for the Perplexed." *Social Studies of Science*, vol. 47, no. 1, 2017, pp. 117-142.

Lynch, Michael P. "Kick This Rock: Climate Change and Our Common Reality." *The New York Times*, June 5, 2017.

Lynch, Michael. "Against Reflexivity as an Academic Virtue and Source of Privileged Knowledge." *Theory, Culture and Society*, vol. 17, no. 3, 2000, pp. 26-54.

Lynch, Michael, Steve Fuller, et al. *Social Studies of Science*, vol. 36, no. 6, 2006, pp. 819-68.

Macintyre, Alasdair. *After Virtue: A Study in Moral Theory*. U of Notre Dame P, 1981.

Mahmood, Saba. *Politics of Piety: The Islamic Revival and the Feminist Subject*. Princeton UP, 2005.

Manda, Neera. *Prophets Looking Backward: Postmodern Critiques of Science and Hindu Nationalism in India*. Rutgers UP, 2003.

Marshall, George. *Don't Even Think About It: Why Our Brains are Wired to Ignore Climate Change*. Bloomsbury Publishing, 2014.

Martin, Wallace. "Criticism and the Academy." *Modernism and the New Criticism*, edited by A. Walton Litz, Louis Menand, and Lawrence Rainey, *The Cambridge History of Literary Criticism*, vol. 7, Cambridge UP, 2000, pp. 267-321.

Maturana, Humberto, and Francisco J. Varela. *Autopoiesis and Cognition: The Realization of the Living*. Dordrecht, Holland, 1980.

Melser, Derek. *The Act of Thinking*. MIT P, 2004.

Meyer, Leonard B. *Emotion and Meaning in Music*. U of Chicago P, 1956.

Miller, Adam S. *Speculative Grace: Bruno Latour and Object-Oriented Theology*. Fordham UP, 2013.

Mirowski, Philip. *Science-Mart: Privatizing American Science*. Harvard UP, 2011.

Mohanty, Satya. *Literary Theory and the Claims of History: Postmodernism, Objectivity, Multicultural Politics*. Cornell UP, 1997.

Moretti, Franco. "Conjectures on World Literature." *New Left Review 1*, 2000, pp. 54–68.

———. *Graphs, Maps, and Trees: Abstract Models for Literary History*. Verso, 2007.

———. "Style, Inc. Reflections on Seven Thousand Titles (British Novels, 1740–1850)." *Critical Inquiry*, vol. 36, no. 1, 2009, pp. 134–58.

Nelson, Robert H. *The New Holy Wars: Economic Religion versus Environmental Religion in Contemporary America*. Penn State UP, 2010.

Nisbett, Richard E. *The Geography of Thought: How Asians and Westerners Think Differently ... and Why*. Free P, 2003.

Noble, David F. *A World Without Women: The Christian Clerical Culture of Western Science*. Knopf, 1992.

Noë, Alva. *Action in Perception*. MIT P, 2004.

Núñez Rafael, and Walter J. Freeman, editors. *Reclaiming Cognition: The Primacy of Action, Intention, and Emotion*. Imprint Academic, 1999.

Oreskes, Naomi and Eric M. Conway. *The Collapse of Western Civilization: A View from the Future*. Columbia UP, 2014.

Oyama, Susan. *The Ontogeny of Information: Developmental Systems and Evolution*. Duke UP, 2000 [originally 1985].

Oyama, Susan, Paul E. Griffiths, and Russell D. Gray, editors. *Cycles of Contingency: Developmental Systems and Evolution*. MIT P, 2001.

Petitot, Jean, Francisco J. Varela, Bernard Pachoud, and Jean-Michel Roy, editors. *Naturalizing Phenomenology: Issues in Contemporary Phenomenology and Cognitive Science*. Stanford UP, 2000.

Pfau, Thomas. *Minding the Modern: Human Agency, Intellectual Traditions, and Responsible Knowledge*. U of Notre Dame P, 2013.

Pickering, Andrew. *Science as Practice and Culture*. U of Chicago P, 1992.

———. *The Mangle of Practice: Time, Agency, and Science*. U of Chicago P, 1995.

Pigliucci, Massimo, and Gerd B. Müller, editors. *Evolution: The Extended Synthesis*. MIT P, 2010.

Popper, Karl R. *Conjectures and Refutations: The Growth of Scientific Knowledge*. Basic, 1962.

Port, Robert F., and Timothy van Gelder, editors. *Mind as Motion: Explorations in the Dynamics of Cognition*. MIT P, 1995.

Purdy, Jedidiah. *After Nature: A Politics for the Anthropocene*. Harvard UP, 2016.

Ransom, John Crowe. "Criticism, Inc." *Virginia Quarterly Review*, vol. 13, no. 4, 1937, pp. 586–603.

Ratzinger, Joseph Cardinal (Pope Benedict XVI). "*Pro Eligendo Romana Pontifice*." April 18, 2005, http://www.vatican.va/gpII/documents/homily-pro-eligendo- pontifice_20050418_en.html.

Richards, I. A. *Practical Criticism: A Study of Literary Judgment*. Kegan Paul, Trench, Trubner, 1929.

Rosenberg, Alex. "Why I Am a Naturalist." *The New York Times*, September 7, 2011, https://opinionator.blogs.nytimes.com/2011/09/17/why-i-am-a-naturalist/.

Rouse, Joseph. "Vampires: Social Constructivism, Realism, and Other Philosophical Undead." *History and Theory*, vol. 41, no. 1, 2002, pp. 60–78.

Saklofske, Jon, Estelle Clements, and Richard Cunningham. "They Have Come, Why Won't We Build It? On the Digital Future of the Humanities." *Digital Humanities Pedagogy: Practices, Principles, and Politics*, edited by Brett D. Hirsch, Open Book, 2012, pp. 311–30.

Saler, Benson. *Conceptualizing Religion: Immanent Anthropologists, Transcendent Natives, and Unbounded Categories*. Brill, 1993.

Schmidgen, Henning. *Bruno Latour in Pieces: An Intellectual Biography*. Translated by Gloria Custance, Fordham UP, 2015.

Schneider, Nathan. "Religion, Science and the Humanities: An Interview with Barbara Herrnstein Smith," https://tif.ssrc.org/2010/06/21/religion-science-and-the-humanities/, June 10, 2010.

Searle, John. "Why Should You Believe It?" *New York Review of Books*, vol. 56, no. 14, September 24, 2009, pp. 88–92.

Shapin, Steven. *The Scientific Life: A Moral History of a Late Modern Vocation*. U of Chicago P, 2008.

Sharma, Kriti. *Interdependence: Biology and Beyond*. Fordham UP, 2015.

Simis, Molly J., Haley Madden, and Michael A. Cacciatore. "The Lure of Rationality: Why does the Deficit Model Persist in Science Communication?" *Public Understanding of Science*, 2016, vol. 25, no. 4, pp. 400–414.

Slingerland, Edward. *What Science Offers the Humanities: Integrating Body and Culture*. Cambridge UP, 2008.

Smith, Barbara Herrnstein. *Contingencies of Value: Alternative Perspectives for Critical Theory*. Harvard UP, 1988.

———. *Belief and Resistance: Dynamics of Contemporary Intellectual Controversy*. Harvard UP, 1997.

———. "Comment." Judith Jarvis Thomson, *Goodness and Advice*, edited by Amy Gutman, Princeton UP, 2001, pp. 132-146.

———. "Reply to an Analytic Philosopher." *The South Atlantic Quarterly*, vol. 101, no.1, 2002, pp. 229-242.

———. *Scandalous Knowledge: Science, Truth and the Human.* U of Edinburgh P, 2005/Duke UP, 2006.

———. "Relativism, Today and Yesterday." *Common Knowledge,* vol. 13, nos. 2–3, 2007, pp. 227-49.

———. *Natural Reflections: Human Cognition at the Nexus of Science and Religion.* Yale UP, 2009.

———. "Science and Religion: Lives and Rocks." *The New York Times,* January 25, 2010, http://opinionator.blogs.nytimes.com/2010/01/25/science-andreligion-lives-and-rocks/.

Smith, Jonathan Z. "Religion, Religions, Religious." *Critical Terms for Religious Study,* edited by Mark C. Taylor, U of Chicago P, 1998, pp. 269–84.

Sokal, Alan, and Claude Bricmont. *Fashionable Nonsense: Postmodern Intellectuals' Abuse of Science.* Picador, 1998.

Stanford, P. Kyle. *Exceeding Our Grasp: Science, History, and the Problem of Unconceived Alternatives.* Oxford UP, 2006.

———. "Naturalism without Scientism." *The Blackwell Companion to Naturalism,* edited by Kelly James Clark, John Wiley & Sons, 2016, pp. 91-108.

Stengers, Isabelle. *In Catastrophic Times: Resisting the Coming Barbarism.* Translated by Andrew Goffey, Open Humanities P, 2015.

Stephenson, Wen. *What We're Fighting for Now Is Each Other: Dispatches from the Front Lines of Climate Justice.* Beacon, 2015.

Stoknes, Per Espen. *What We Think About (When We Try Not to Think About) Global Warming: Toward a New Psychology of Climate Action.* Chelsea Green Publishing, 2015.

Taylor, Charles. *A Secular Age.* Harvard UP, 2007.

Thelen, Esther, and L. B. Smith. *A Dynamic Systems Approach to the Development of Cognition and Action.* MIT P, 1994.

Thompson, Evan. *Mind in Life: Biology, Phenomenology, and the Sciences of Mind.* Harvard UP, 2010.

Thomson, Judith Jarvis. *Goodness and Advice.* Edited by Amy Gutman, Princeton UP, 2001.

Tooby, John, and Leda Cosmides. "The Psychological Foundations of Culture." *The Adapted Mind: Evolutionary Psychology and the Generation of Culture,* edited by Jerome H. Barkow, Leda Cosmides, and John Tooby, Oxford UP, 1992, pp. 19-136.

Trumpener, Kate. "Paratext and Genre System: A Response to Franco Moretti." *Critical Inquiry*, vol. 36, no. 1, 2009, pp. 159–71.

van Gelder, Tim, and Robert F. Port. "It's About Time: An Overview of the Dynamical Approach to Cognition." *Mind as Motion: Explorations in the Dynamics of Cognition*, edited by Tim van Gelder and Robert F. Port, MIT P, 1995, pp. 1-44.

Varela, Francisco J., Evan Thompson, and Eleanor Rosch. *The Embodied Mind: Cognitive Science and Human Experience*. MIT P, 1991.

Vermeule, Blakey. "Wit and Poetry and Pope, or The Handicap Principle." *Critical Inquiry*, vol. 38, no. 2, 2012, pp. 426–30.

Veyne, Paul. *Did the Greeks Believe in Their Myths? An Essay on the Constitutive Imagination*. U of Chicago P, 1988.

Vivieros de Castro, Eduardo. *Cannibal Metaphysics*. U of Minnesota P, 2015.

Wilkens, Matthew. "Canons, Close Reading, and the Evolution of Method." *Debates in the Digital Humanities*, edited by Matthew K. Gold, U of Minnesota P, 2011, pp. 249–58.

Williams, Jeffrey J. "The New Modesty in Literary Criticism." *The Chronicle of Higher Education*, January 5, 2015.

Wilson, Edward O. *Consilience: The Unity of Knowledge*. Vintage, 1998.

Wilson, Edward O., and Stephen R. Kellert, editors. *The Biophilia Hypothesis*. Island P, 1993.

Wilson, Elizabeth A. *Psychosomatic: Feminism and the Neurological Body*. Duke UP, 2004.

Wolfram, Stephen. "Injecting Computation Everywhere." blog.stephenwolfram.com/2014/03/injecting-computation-everywhere-a-sxsw-update/. Accessed 4 September 2014.

Zuiderent-Jerak, Teun, and Casper Bruun Jensen. "Unpacking 'Intervention' in Science and Technology Studies." *Science as Culture*, vol. 16, no. 3, 2007, pp. 227–35.

Zunshine, Lisa, editor. *Introduction to Cognitive Cultural Studies*. Johns Hopkins UP, 2010.

Acknowledgements

Several chapters draw on previously published texts. All have been edited and to various extents revised for this volume.

Chapter 2, "The Chimera of Relativism: A Tragicomedy," was originally published in *Common Knowledge*, vol. 17, no.1, 2011, pp. 13-26. © Duke University Press. Reprinted with permission. It is based on a talk delivered under the title "Angles on Relativism: Historical, Anthropological, Political, and Philosophical" at a colloquium on Comparative Relativism at the IT University of Copenhagen in September 2009.

Chapter 3, "Religion, Science, and the Humanities: An Interview with Barbara Herrnstein Smith," was initially posted on June 21, 2010 on the Social Sciences Research Council blog *The Immanent Frame*. It is archived there as https://tif.ssrc.org/2010/06/21/religion-science-and-the-humanities/.

Chapter 4 is based on a talk delivered at a conference on Recomposing the Humanities—with Bruno Latour at the University of Virginia in September 2015. The text, originally titled "Anthropotheology: Latour Speaking Religiously," was published in *New Literary History* 47.2-3 (2016), 331-351. Copyright © 2016 The University of Virginia. Reprinted with permission by Johns Hopkins University Press.

Chapter 5, "What Was 'Close Reading'?: A Century of Method in Literary Studies," was originally published in *The Minnesota Review*, vol. 87, 2016, pp. 57-75, © Duke University Press. Reprinted with permission. It is based on a talk originally delivered at a Digital Humanities Workshop on Method at the Heyman Center, Columbia University, New York, in May 2015.

Earlier versions of Chapter 6 were delivered under the title "Dis/Integration: The New Interdisciplinarity" as the Stephen Straker Lecture at the University of British Columbia in October 2013 and as the Lahey Lecture at Concordia University, Montreal, in September 2014. "Scientizing the Humanities: Shifts, Collisions, Negotiations" was originally published in *Common Knowledge*, vol. 22, no. 3, 2016, pp. 353-372, © Duke University Press. Reprinted with permission.

Chapter 7, "Perplexing Realities: Practicing Relativism in the Anthropocene," is a revised version of a talk delivered at a symposium on Climate Realism sponsored by McGill University in Montreal in March 2017.

My thanks to Ulrik Ekman, Casper Bruun Jensen, Mi Gyung (Mimi) Kim, Kurt Kohn, Claudia Koonz, Susan Oyama, Deirdre M. Smith, and E. Roy Weintraub for evocative conversations regarding the topics treated here and for valuable readings of the drafts of individual chapters. Much is owed to Tom Cohen and Claire Colebrook for their contributions to the shaping of this volume. Special thanks to Stephen Barber for serving, these many years, as personal first responder to thoughts and prose.

www.ingramcontent.com/pod-product-compliance
Lightning Source LLC
Chambersburg PA
CBHW031149160426
43193CB00008B/309